Hyperparameter Tuning with Python

Boost your machine learning model's performance
via hyperparameter tuning

Louis Owen

Pack**t>**

BIRMINGHAM—MUMBAI

Hyperparameter Tuning with Python

Group Product Manager: Gebin George
Publishing Product Manager: Dinesh Chaudhary
Senior Editor: David Sugarman
Technical Editor: Devanshi Ayare
Copy Editor: Safis Editing
Project Coordinator: Farheen Fathima
Proofreader: Safis Editing
Indexer: Pratik Shirodkhar
Production Designer: Ponraj Dhandapani
Marketing Coordinator: Shifa Ansari and Abeer Riyaz Dawe

First published: July 2022

Production reference: 1280722

Published by Packt Publishing Ltd.
Livery Place
35 Livery Street
Birmingham
B3 2PB, UK.

ISBN 978-1-80323-587-5

www.packt.com

To Mom and Dad, thanks for everything!

– Louis

Contributors

About the author

Louis Owen is a data scientist/AI engineer from Indonesia who is always hungry for new knowledge. Throughout his career journey, he has worked in various fields of industry, including NGOs, e-commerce, conversational AI, OTA, Smart City, and FinTech. Outside of work, he loves to spend his time helping data science enthusiasts to become data scientists, either through his articles or through mentoring sessions. He also loves to spend his spare time doing his hobbies: watching movies and conducting side projects. Finally, Louis loves to meet new friends! So, please feel free to reach out to him on LinkedIn if you have any topics to be discussed.

About the reviewer

Jamshaid Sohail is passionate about data science, machine learning, computer vision, and natural language processing and has more than 2 years of experience in the industry. He has worked at a Silicon Valley-based start-up named FunnelBeam, the founders of which are from Stanford University, as a data scientist. Currently, he is working as a data scientist at Systems Limited. He has completed over 66 online courses from different platforms. He authored the book Data Wrangling with Python 3.X for Packt Publishing and has reviewed multiple books and courses. He is also developing a comprehensive course on data science at Educative and is in the process of writing books for multiple publishers.

Table of Contents

Section 2: The Implementation

7

Hyperparameter Tuning via Scikit

8

Hyperparameter Tuning via Hyperopt

9

Hyperparameter Tuning via Optuna

Section 3: Putting Things into Practice

12

Introducing Hyperparameter Tuning Decision Map

13

Tracking Hyperparameter Tuning Experiments

14

Conclusions and Next Steps

Index

Other Books You May Enjoy

Preface

Hyperparameters are an important element in building useful machine learning models. This book curates numerous hyperparameter tuning methods for Python, one of the most popular coding languages for machine learning. Alongside in-depth explanations of how each method works, you will use a decision map that can help you identify the best tuning method for your requirements.

We will start the book with an introduction to hyperparameter tuning and explain why it's important. You'll learn the best methods for hyperparameter tuning for a variety of use cases and a specific algorithm type. The book will not only cover the usual grid or random search but also other powerful underdog methods. Individual chapters are dedicated to giving full attention to the three main groups of hyperparameter tuning methods: exhaustive search, heuristic search, Bayesian optimization, and multi-fidelity optimization.

Later in the book, you will learn about top frameworks such as scikit-learn, Hyperopt, Optuna, NNI, and DEAP to implement hyperparameter tuning. Finally, we will cover hyperparameters of popular algorithms and best practices that will help you efficiently tune your hyperparameters.

By the end of the book, you will have the skills you need to take full control over your machine learning models and get the best models for the best results.

Who this book is for

The book is intended for data scientists and Machine Learning engineers who are working with Python and want to further boost their ML model's performance by utilizing the appropriate hyperparameter tuning method. You will need to have a basic understanding of ML and how to code in Python but will require no prior knowledge of hyperparameter tuning in Python.

What this book covers

Chapter 1, *Evaluating Machine Learning Models*, covers all the important things we need to know when it comes to evaluating ML models, including the concept of overfitting, the idea of splitting data into several parts, a comparison between the random and stratified split, and numerous methods on how to split the data.

Chapter 2, *Introducing Hyperparameter Tuning*, introduces the concept of hyperparameter tuning, starting from the definition and moving on to the goal, several misconceptions, and distributions of hyperparameters.

Chapter 3, Exploring Exhaustive Search, explores each method that belongs to the first out of four groups of hyperparameter tuning, along with the pros and cons. There will be both high-level and detailed explanations for each of the methods. The high-level explanation will use a visualization strategy to help you understand more easily, while the detailed explanation will bring the math to the table.

Chapter 4, Exploring Bayesian Optimization, explores each method that belongs to the second out of four groups of hyperparameter tuning, along with the pros and cons. There will also be both high-level and detailed explanations for each of the methods.

Chapter 5, Exploring Heuristic Search, explores each method that belongs to the third out of four groups of hyperparameter tuning, along with the pros and cons. There will also be both high-level and detailed explanations for each of the methods.

Chapter 6, Exploring Multi-Fidelity Optimization, explores each method that belongs to the fourth out of four groups of hyperparameter tuning, along with the pros and cons. There will also be both high-level and detailed explanations for each of the methods.

Chapter 7, Hyperparameter Tuning via Scikit, covers all the important things about scikit-learn, scikit-optimize, and scikit-hyperband, along with how to utilize each of them to perform hyperparameter tuning.

Chapter 8, Hyperparameter Tuning via Hyperopt, introduces the Hyperopt package, starting from its capabilities and limitations, how to utilize it to perform hyperparameter tuning, and all the other important things you need to know about it.

Chapter 9, Hyperparameter Tuning via Optuna, introduces the Optuna package, starting from its numerous features, how to utilize it to perform hyperparameter tuning, and all the other important things you need to know about it.

Chapter 10, Advanced Hyperparameter Tuning with DEAP and Microsoft NNI, shows how to perform hyperparameter tuning using both the DEAP and Microsoft NNI packages, starting from getting ourselves familiar with the packages and moving on to the important modules and parameters we need to be aware of.

Chapter 11, Understanding Hyperparameters of Popular Algorithms, explores the hyperparameters of several popular ML algorithms. There will be a broad explanation for each of the algorithms, including (but not limited to) the definition of each hyperparameter, what will be impacted when the value of each hyperparameter is changed, and the priority list of hyperparameters based on the impact.

Chapter 12, Introducing Hyperparameter Tuning Decision Map, introduces the **Hyperparameter Tuning Decision Map (HTDM)**, which summarizes all of the discussed hyperparameter tuning methods as a simple decision map based on six aspects. There will be also three study cases that show how to utilize the HTDM in practice.

Chapter 13, Tracking Hyperparameter Tuning Experiments, covers the importance of tracking hyperparameter tuning experiments, along with the usual practices. You will also be introduced to several open source packages that are available and learn how to utilize each of them in practice.

Chapter 14, Conclusions and Next Steps, summarizes all the important lessons learned in the previous chapters, and also introduces you to several topics or implementations that you may benefit from that we have not covered in detail in this book.

To get the most out of this book

You will also need Python version 3.7 (or above) installed on your computer, along with the related packages mentioned in the *Technical requirements* section of each chapter.

It is worth noting that there is a conflicting version requirement for the Hyperopt package in *Chapter 8, Hyperparameter Tuning via Hyperopt*, and *Chapter 10, Advanced Hyperparameter Tuning with DEAP and Microsoft NNI*. You need to install version 0.2.7 for *Chapter 8, Hyperparameter Tuning via Hyperopt*, and version 0.1.2 for *Chapter 10, Advanced Hyperparameter Tuning with DEAP and Microsoft NNI*.

It is also worth noting that the `HyperBand` implementation used in *Chapter 7, Hyperparameter Tuning via Scikit*, is the modified version of the scikit-hyperband package. You can utilize the modified version by cloning the GitHub repository (a link is available in the next section) and looking in a folder named `hyperband`.

If you are using the digital version of this book, we advise you to type the code yourself or access the code from the book's GitHub repository (a link is available in the next section). Doing so will help you avoid any potential errors related to the copying and pasting of code.

To understand all contents in this book, you will need to have a basic understanding of ML and how to code in Python but will require no prior knowledge of hyperparameter tuning in Python. At the end of this book, you will also be introduced to several topics or implementations that you may benefit from which we have not covered yet in this book.

Download the example code files

You can download the example code files for this book from GitHub at `https://github.com/PacktPublishing/Hyperparameter-Tuning-with-Python`. If there's an update to the code, it will be updated in the GitHub repository.

We also have other code bundles from our rich catalog of books and videos available at `https://github.com/PacktPublishing/`. Check them out!

Download the color images

We also provide a PDF file that has color images of the screenshots and diagrams used in this book. You can download it here: `https://packt.link/ExcbH`.

Conventions used

There are a number of text conventions used throughout this book.

`Code in text`: Indicates code words in text, database table names, folder names, filenames, file extensions, pathnames, dummy URLs, user input, and Twitter handles. Here is an example: As for `criterion` and `max_depth`, we are still using the same configuration as the previous search space.

A block of code is set as follows:

```
for n_est in n_estimators:
        for crit in criterion:
        for m_depth in max_depth:
        #perform cross-validation here
```

> **Tips or Important Notes**
> Appear like this.

Get in touch

Feedback from our readers is always welcome.

General feedback: If you have questions about any aspect of this book, email us at `customercare@packtpub.com` and mention the book title in the subject of your message.

Errata: Although we have taken every care to ensure the accuracy of our content, mistakes do happen. If you have found a mistake in this book, we would be grateful if you would report this to us. Please visit `www.packtpub.com/support/errata` and fill in the form.

Piracy: If you come across any illegal copies of our works in any form on the internet, we would be grateful if you would provide us with the location address or website name. Please contact us at `copyright@packt.com` with a link to the material.

If you are interested in becoming an author: If there is a topic that you have expertise in and you are interested in either writing or contributing to a book, please visit authors.packtpub.com.

Share Your Thoughts

Once you've read *Hyperparameter Tuning with Python*, we'd love to hear your thoughts! Scan the QR code below to go straight to the Amazon review page for this book and share your feedback.

https://packt.link/r/1-803-23587-X

Your review is important to us and the tech community and will help us make sure we're delivering excellent quality content.

Section 1: The Methods

This initial section covers concepts and theories you need to know before performing hyperparameter tuning experiments.

This section includes the following chapters:

1

Evaluating Machine Learning Models

Machine Learning (ML) models need to be thoroughly evaluated to ensure they will work in production. We have to ensure the model is not *memorizing* the training data and also ensure it learns enough from the given training data. Choosing the appropriate evaluation method is also critical when we want to perform hyperparameter tuning at a later stage.

In this chapter, we'll learn about all the important things we need to know when it comes to evaluating ML models. First, we need to understand the concept of overfitting. Then, we will look at the idea of splitting data into train, validation, and test sets. Additionally, we'll learn about the difference between random and stratified splits and when to use each of them.

We'll discuss the concept of cross-validation and its numerous variations of strategy: k-fold repeated k-fold, **Leave One Out** (**LOO**), **Leave P Out** (**LPO**), and a specific strategy when dealing with time-series data, called time-series cross-validation. We'll also learn how to implement each of the evaluation strategies using the Scikit-Learn package.

By the end of this chapter, you will have a good understanding of why choosing a proper evaluation strategy is critical in the ML model development life cycle. Also, you will be aware of numerous evaluation strategies and will be able to choose the most appropriate one for your situation. Furthermore, you will also be able to implement each of the evaluation strategies using the Scikit-Learn package.

In this chapter, we're going to cover the following main topics:

- Understanding the concept of overfitting
- Creating training, validation, and test sets
- Exploring random and stratified split
- Discovering k-fold cross-validation
- Discovering repeated k-fold cross-validation

- Discovering LOO cross-validation

- Discovering LPO cross-validation

- Discovering time-series cross-validation

Technical requirements

We will learn how to implement each of the evaluation strategies using the Scikit-Learn package. To ensure that you can reproduce the code examples in this chapter, you will need the following:

- Python 3 (version 3.7 or above)

- The pandas package installed (version 1.3.4 or above)

- The Scikit-Learn package installed (version 1.0.1 or above)

All of the code examples for this chapter can be found on GitHub at `https://github.com/PacktPublishing/Hyperparameter-Tuning-with-Python/blob/main/01_Evaluating-Machine-Learning-Models.ipynb`.

Understanding the concept of overfitting

Overfitting occurs when the trained ML model learns too much from the given training data. In this situation, the trained model successfully gets a high evaluation score on the training data but a far lower score on new, unseen data. In other words, the trained ML model fails to generalize the knowledge learned from the training data to the unseen data.

So, how exactly does the trained ML model get decent performance on the training data but fail to give a reasonable performance on unseen data? Well, that happens when the model *tries too hard* to achieve high performance on the training data and has picked up knowledge that is only applicable to that specific training data. Of course, this will negatively impact the model's ability to generalize, which results in bad performance when the model is evaluated on unseen data.

To detect whether our trained ML model faces an overfitting issue, we can monitor the performance of our model on the training data versus unseen data. Performance can be defined as the loss value of our model or metrics that we care about, for example, accuracy, precision, and the mean absolute error. If the performance of the training data keeps getting better, while the performance on the unseen data starts to become stagnant or even gets worse, then this is a sign of an overfitting issue (see *Figure 1.1*):

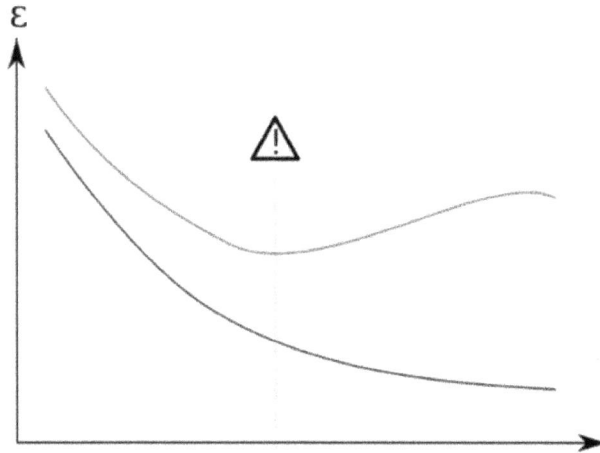

Figure 1.1 – The model's performance on training data versus unseen data (overfitting)

> **Note**
>
> The preceding diagram image has been reproduced according to the license specified: `https://commons.wikimedia.org/wiki/File:Overfitting_svg.svg`.

Now that you are aware of the overfitting problem, we need to learn how to prevent this from happening in our ML development life cycle. We will discuss this in the following sections.

Creating training, validation, and test sets

We understand that overfitting can be detected by monitoring the model's performance on the training data versus the unseen data, but what exactly is unseen data? Is it just random data that has not yet been seen by the model during the training phase?

Unseen data is a portion of our original complete data that was not seen by the model during the training phase. We usually refer to this unseen data as the **test set**. Let's imagine you have 100,000 samples of data, to begin with; you can take out a portion of the data, let's say 10% of it, to become the test set. So, now we have 90,000 samples as the training set and 10,000 samples as the testing set.

However, it is better to not just split our original data into train and test sets but also into a **validation set**, *especially when we want to perform hyperparameter tuning* on our model. Let's say that out of 100,000 original samples, we held out 10% of it to become the validation set and another 10% to become the test set. Therefore, we will have 80,000 samples as the train set, 10,000 samples as the validation set, and 10,000 samples as the test set.

You might be wondering why do we need a validation set apart from the test set. Actually, we do not need it if we do not want to perform hyperparameter tuning or any other *model-centric* approaches. This is because the purpose of having a validation set is to have an unbiased evaluation of the test set using the final version of the trained model.

A validation set can help us to get an unbiased evaluation of the test set because we only incorporate the validation set during the hyperparameter tuning phase. Once we finish the hyperparameter tuning phase and get the final model configuration, we can then evaluate our model on the purely unseen data, which is called the test set.

> **Important Note**
>
> If you are going to perform any data preprocessing steps (for example, missing value imputation, feature engineering, standardization, label encoding, and more), you have to build the function based on the train set and then apply it to the validation and test set. Do *not* perform those data preprocessing steps on the full original data (before data splitting). That's because it might lead to a **data leakage** problem.

There is no specific rule when it comes to choosing the proportions for each of the train, validation, and test sets. You have to choose the split proportion by yourself based on the condition you are faced with. However, the common splitting proportion used by the data science community is 8:2 or 9:1 for the train set and the validation and test set, respectively. Usually, the validation and test set will have a proportion of 1:1. Therefore, the common splitting proportion is 8:1:1 or 9:0.5:0.5 for the train, validation, and test sets, respectively.

Now that we are aware of the train, validation, and test set concept, we need to learn how to build those sets. Do we just randomly split our original data into three sets? Or can we also apply some predefined rules? In the next section, we will explore this topic in more detail.

Exploring random and stratified splits

The most straightforward way (but *not entirely a correct way*) to split our original full data into train, validation, and test sets is by choosing the proportions for each set and then directly splitting them into three sets based on the order of the index.

For instance, the original full data has 100,000 samples, and we want to split this into train, validation, and test sets with a proportion of 8:1:1. Then, the training set will be the samples from index 1 until 80,000. The validation and test set will be the index from 81,000 until 90,000 and 91,000 until 100,000, respectively.

So, what's wrong with that approach? There is nothing wrong with that approach *as long as the original full data is shuffled*. It might cause a problem when there is some kind of pattern between the indices of the samples.

For instance, we have data consisting of 10,000 samples and 3 columns. The first and second columns contain weight and height information, respectively. The third column contains the "weight status" class (for example, underweight, normal weight, overweight, and obesity). Our task is to build an ML classifier model to predict what the "weight status" class of a person is, given their weight and height. It is not impossible for the data to be given to us in the condition that it was ordered based on the third column. So, the first 80,000 rows only consist of the underweight and normal weight classes. In comparison, the overweight and obesity classes are only located in the last 20,000 rows. If this is the case, and we apply the data splitting logic from earlier, then there is no way our classifier can predict a new person has the overweight or obesity "weight status" classes. Why? Because our classifier has never seen those classes before during the training phase!

Therefore, it is very important to ensure the original full data is shuffled in the first place, and essentially, this is what we mean by the random split. **Random split** works by first shuffling the original full data and then splitting it into the train, validation, and test sets based on the order of the index.

There is also another splitting logic called the **stratified split**. This logic ensures that the train, validation, and test set will get *a similar proportion number of samples for each target class* found in the original full data.

Using the same "weight status" class prediction case example, let's say that we found that the proportion of each class in the full original data is 3:5:1.5:0.5 for underweight, normal weight, overweight, and obese, respectively. The stratified split logic will ensure that we can find a similar proportion of those classes in the train, validation, and test sets. So, out of 80,000 samples of the train set, around 24,000 samples are in the underweight class, around 40,000 samples are in the normal weight class, around 12,000 samples are overweight, and around 4,000 samples are in the obesity class. This will also be applied to the validation and test set.

The remaining question is understanding *when it is the right time* to use the random split/stratified split logic. Often, the stratified split logic is used when we are faced with an imbalanced class problem. However, it is also often used when we want to make sure that we have a similar proportion of samples based on a specific variable (not necessarily the target class). If you are not faced with this kind of situation, then the random split is the go-to logic that you can always choose.

To implement both of the data splitting logics, you can write the code by yourself from scratch or utilize the well-known package called **Scikit-Learn**. The following is an example to perform a random split with a proportion of 8:1:1:

```
from sklearn.model_selection import train_test_split
df_train, df_unseen = train_test_split(df, test_size=0.2,
random_state=0)
df_val, df_test = train_test_split(df_unseen, test_size=0.5,
random_state=0)
```

The df variable is our complete original data that was stored in the Pandas DataFrame object. The train_test_split function splits the Pandas DataFrame, array, or matrix into shuffled train and test sets. In lines 2–3, first, we split the original full data into df_train and df_unseen with a proportion of 8:2, as specified by the test_size argument. Then, we split df_unseen into df_val and df_test with a proportion of 1:1.

To perform the *stratify split* logic, you can just add the stratify argument to the train_test_split function and fill it with the target array:

```
df_train, df_unseen = train_test_split(df, test_size=0.2,
random_state=0, stratify=df['class'])
df_val, df_test = train_test_split(df_unseen, test_size=0.5,
random_state=0, stratify=df_unseen['class'])
```

The stratify argument will ensure the data is split in the stratified fashion based on the given target array.

In this section, we have learned the importance of shuffling the original full data before performing data splitting and also understand the difference between the random and stratified split, as well as when to use each of them. In the next section, we will start learning variations of the data splitting strategies and how to implement each of them using the Scikit-learn package.

Discovering k-fold cross-validation

Cross-validation is a way to evaluate our ML model by performing *multiple evaluations* on our original full data via a resampling procedure. This is a variation from the vanilla train-validation-test split that we learned about in previous sections. Additionally, the concept of random and stratified splits can be applied in cross-validation.

In cross-validation, we perform multiple splits for the *train and validation* sets, where each split is usually referred to **Fold**. What about the test set? Well, it still acts as the purely unseen data where we can test the final model configuration on it. Therefore, in the beginning, it is only separated once from the train and validation set.

There are several variations of the cross-validation strategy. The first one is called **k-fold cross-validation**. It works by performing k times of training and evaluation with a proportion of (k-1):1 for the train and validation set, respectively, in each fold. To have a clearer understanding of k-fold cross-validation, please refer to *Figure 1.2*:

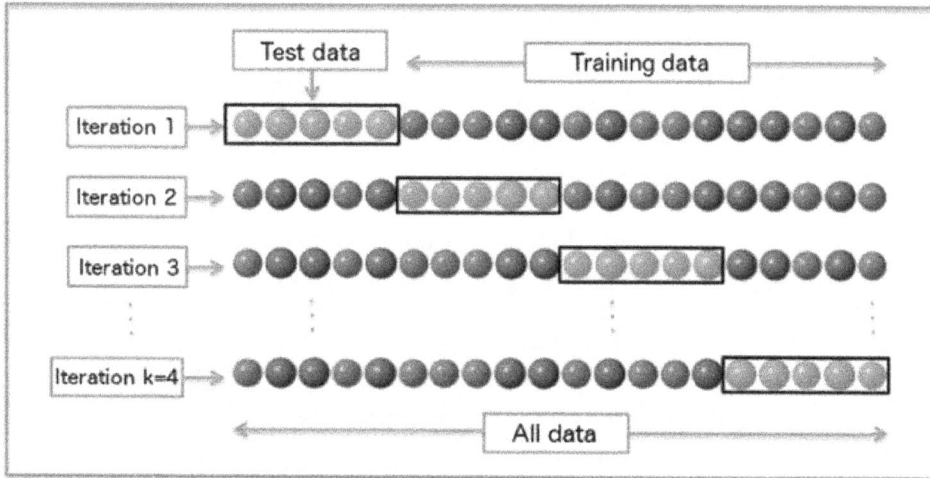

Figure 1.2 – K-fold cross-validation

> **Note**
>
> The preceding diagram has been reproduced according to the license specified: `https://commons.wikimedia.org/wiki/File:K-fold_cross_validation.jpg`.

For instance, let's choose k = 4 to match the illustration in *Figure 1.2*. The green and red balls correspond to the target class, where, in this case, we only have two target classes. The data is shuffled beforehand, which can be seen from the absence of a pattern of green and red balls. It is also worth mentioning that the *shuffling was previously only done once*. That's why the order of green and red balls is always the same for each iteration (fold). The black box in each fold corresponds to the validation set (the test data is in the illustration).

As you can see in *Figure 1.2*, the proportion of the training set versus the validation set is (k-1):1, or in this case, 3:1. During each fold, the model will be trained on the train set and evaluated on the validation set. Notice that the training and validation sets are *different across each fold*. The final evaluation score can be calculated by taking the average score of all of the folds.

In summary, k-fold cross-validation works as follows:

1. Shuffling the original full data
2. Holding out the test data
3. Performing the k-fold multiple evaluation strategy on the rest of the original full data
4. Calculating the final evaluation score by taking the average score of all of the folds
5. Evaluating the test data using the final model configuration

You might ask why do we need to perform cross-validation in the first place? Why is the vanilla train-validation-test splitting strategy not enough? There are several reasons why we need to apply the cross-validation strategy:

- Having only a small amount of training data.

- To get a more confident conclusion from the evaluation performance.

- To get a clearer picture of our model's learning ability and/or the complexity of the given data.

The first and second reasons are quite straightforward. The third reason is more interesting and should be discussed. How can cross-validation help us to get a better idea about our model's learning ability and/or the data complexity? Well, this happens when the variation of evaluation scores from each fold is quite big. For instance, out of 4 folds, we get accuracy scores of 45%, 82%, 64%, and 98%. This scenario should trigger our curiosity: what is wrong with our model and/or data? It could be that the data is too hard to learn and/or our model can't learn properly.

The following is the syntax to perform k-fold cross-validation via the Scikit-Learn package:

```
From sklearn.model_selection import train_test_split, Kfold
df_cv, df_test = train_test_split(df, test_size=0.2, random_
state=0)
kf = Kfold(n_splits=4)
for train_index, val_index in kf.split(df_cv):
df_train, df_val = df_cv.iloc[train_index], df_cv.iloc[val_
index]
#perform training or hyperparameter tuning here
```

Notice that, first, we hold out the test set and only work with df_cv when performing the k-fold cross-validation. By default, the Kfold function will disable the shuffling procedure. However, this is not a problem for us since the data has already shuffled beforehand when we called the train_test_split function. If you want to run the shuffling procedure again, you can pass shuffle=True in the Kfold function.

Here is another example if you are interested in learning how to apply the concept of stratifying splits in k-fold cross-validation:

```
From sklearn.model_selection import train_test_split,
StratifiedKFold
df_cv, df_test = train_test_split(df, test_size=0.2, random_
state=0, stratify=df['class'])
skf = StratifiedKFold(n_splits=4)
```

```
for train_index, val_index in skf.split(df_cv, df_cv['class']):
df_train, df_val = df_cv.iloc[train_index], df_cv.iloc[val_
index]
#perform training or hyperparameter tuning here
```

The only difference is to import `StratifiedKFold` instead of the `Kfold` function and add the array of target variables, which will be used to split the data in a stratified fashion.

In this section, you have learned what cross-validation is, when the right time is to perform cross-validation, and the first (and the *most widely used*) cross-validation strategy variation, which is called k-fold cross-validation. In the subsequent sections, we will also learn other variations of cross-validation and how to implement them using the Scikit-Learn package.

Discovering repeated k-fold cross-validation

Repeated k-fold cross-validation involves simply performing the k-fold cross-validation repeatedly, *N* times, with different randomizations in each repetition. The final evaluation score is the average of all scores from all folds of each repetition. This strategy will increase our confidence in our model.

So, why repeat the k-fold cross-validation? Why don't we just increase the value of k in k-fold? Surely, increasing the value of k will reduce the bias of our model's estimated performance. However, increasing the value of k will increase the variation, especially when we have a small number of samples. Therefore, usually, repeating the k-folds is a better way to gain higher confidence in our model's estimated performance. Of course, this comes with a drawback, which is the increase in computation time.

To implement this strategy, we can simply perform a manual for-loop, where we apply the k-fold cross-validation strategy to each loop. Fortunately, the Scikit-Learn package provide us with a specific function in which to implement this strategy:

```
from sklearn.model_selection import train_test_split,
RepeatedKFold
df_cv, df_test = train_test_split(df, test_size=0.2, random_
state=0)
rkf = RepeatedKFold(n_splits=4, n_repeats=3, random_state=0)
for train_index, val_index in rkf.split(df_cv):
df_train, df_val = df_cv.iloc[train_index], df_cv.iloc[val_
index]
#perform training or hyperparameter tuning here
```

Choosing n_splits=4 and n_repeats=3 means that we will have 12 different train and validation sets. The final evaluation score is then just the average of all 12 scores. As you might expect, there is also a dedicated function to implement the repeated k-fold in a stratified fashion:

```
from sklearn.model_selection import train_test_split,
RepeatedStratifiedKFold
df_cv, df_test = train_test_split(df, test_size=0.2, random_
state=0, stratify=df['class'])
rskf = RepeatedStratifiedKFold(n_splits=4, n_repeats=3, random_
state=0)
for train_index, val_index in rskf.split(df_cv, df_
cv['class']):
df_train, df_val = df_cv.iloc[train_index], df_cv.iloc[val_index]
#perform training or hyperparameter tuning here
```

The RepeatedStratifiedKFold function will perform stratified k-fold cross-validation repeatedly, n_repeats times.

Now that you have learned another variation of the cross-validation strategy, called repeated k-fold cross-validation, let's learn about the other variations next.

Discovering Leave-One-Out cross-validation

Essentially, **Leave One Out (LOO) cross-validation** is just k-fold cross-validation where k = n, where n is the number of samples. This means there are n-1 samples for the training set and 1 sample for the validation set in each fold (see *Figure 1.3*). Undoubtedly, this is a very *computationally expensive* strategy and will result in a *very high variance* evaluation score estimator:

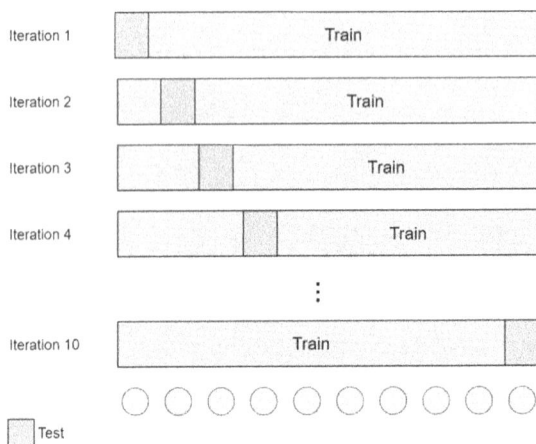

Figure 1.3 – LOO cross-validation

So, when is LOO preferred over k-fold cross-validation? Well, LOO works best when you have a very small dataset. It is also good to choose LOO over k-fold if you prefer the high confidence of the model's performance estimation over the computational cost limitation.

Implementing this strategy from scratch is actually very simple. We just need to loop through each of the indexes of data and do some data manipulation. However, the Scikit-Learn package also provides the implementation for LOO, which we can use:

```
from sklearn.model_selection import train_test_split,
LeaveOneOut
df_cv, df_test = train_test_split(df, test_size=0.2, random_
state=0)
loo = LeaveOneOut()
for train_index, val_index in loo.split(df_cv):
df_train, df_val = df_cv.iloc[train_index], df_cv.iloc[val_
index]
#perform training or hyperparameter tuning here
```

Notice that there is no argument provided in the LeaveOneOut function since this strategy is very straightforward and involves no stochastic procedure. There is also no stratified version of the LOO since the validation set will always contain one sample.

Now that you are aware of the concept of LOO, in the next section, we will learn about a slight variation of LOO.

Discovering LPO cross-validation

LPO cross-validation is a variation of the LOO cross-validation strategy, where the validation set in each fold contains p samples instead of only 1 sample. Similar to LOO, this strategy will ensure that we get all possible combinations of train-validation pairs. To be more precise, there will be $\mathbb{C}_p^n$ number of folds assuming there are n samples on our data. For example, there will be $\mathbb{C}_5^{30}$ or 142,506 folds if we want to perform Leave-5-Out cross-validation on data that has 50 samples.

LPO is suitable when you have a small number of samples and want to get even higher confidence in the model's estimated performance compared to the LOO method. LPO will result in an exploding number of folds when you have a large number of samples.

This strategy is a bit different from k-fold or LOO in terms of the overlapping between the validation sets. For P > 1, LPO will result in overlapping validation sets, while k-fold and LOO will always result in non-overlapping validation sets. Also, note that LPO is different from k-fold with K = N // P since k-fold will always create non-overlapping validation sets, but not with the LPO strategy:

```
from sklearn.model_selection import train_test_split, LeavePOut
df_cv, df_test = train_test_split(df, test_size=0.2, random_state=0)
lpo = LeavePOut(p=2)
for train_index, val_index in lpo.split(df_cv):
df_train, df_val = df_cv.iloc[train_index], df_cv.iloc[val_index]
#perform training or hyperparameter tuning here
```

Unlike LOO, we have to provide the p argument to LPO, which refers to the p values in LPO.

In this section, we have learned about the variations of the LOO cross-validation strategy. In the next section, we will learn how to perform cross-validation on time-series data.

Discovering time-series cross-validation

Time-series data has a unique characteristic in nature. Unlike "normal" data, which is assumed to be **independent and identically distributed** (IID), time-series data does not follow that assumption. In fact, each sample is dependent on previous samples, meaning changing the order of the samples will result in different data interpretations.

Several examples of time-series data are listed as follows:

- Daily stock market price
- Hourly temperature data
- Minute-by-minute web page clicks count

There will be a **look-ahead bias** if we apply previous cross-validation strategies (for example, k-fold or random or stratified splits) to time-series data. Look-ahead bias happens when we use the future value of the data that is supposedly not available for the current time of the simulation.

For instance, we are working with hourly temperature data. We want to predict what the temperature will be in 2 hours, but we use the temperature value of the next hour or the next 3 hours, which is supposedly not available yet. This kind of bias will happen easily if we apply the previous cross-validation strategies since those strategies are designed to work well only on IID distribution.

Time-series cross-validation is the cross-validation strategy that is specifically designed to handle time-series data. It works similarly to k-fold in terms of accepting the predefined values of folds, which then generates k test sets. The difference is that the *data is not shuffled in the first place*, and the training set in the next iteration is the *superset* of the one in the previous iteration, meaning the training set keeps getting bigger over the number of iterations. Once we finish with the cross-validation and get the final model configuration, we can then test our final model on the test data (see *Figure 1.4*):

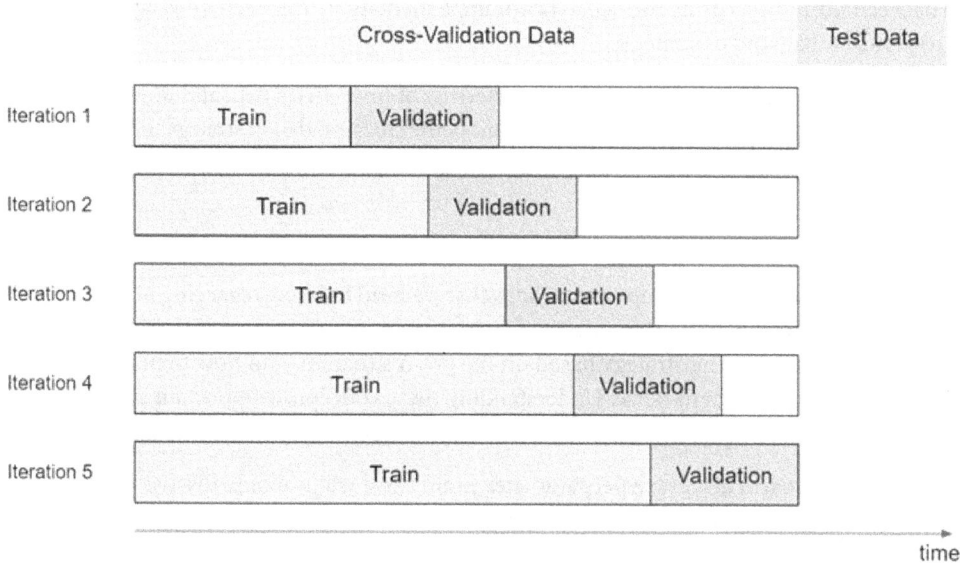

Figure 1.4 – Time-series cross-validation

Also, the Scikit-Learn package provides us with a nice implementation of this strategy:

```
from sklearn.model_selection import train_test_split,
TimeSeriesSplit
df_cv, df_test = train_test_split(df, test_size=0.2, random_
state=0, shuffle=False)
tscv = TimeSeriesSplit(n_splits=5)
for train_index, val_index in tscv.split(df_cv):
df_train, df_val = df_cv.iloc[train_index], df_cv.iloc[val_
index]
#perform training or hyperparameter tuning here
```

Providing n_splits=5 will ensure that there are five test sets generated. It is worth noting that, by default, the train set will have the size of $i \cdot n_{samples}\ /\ /\ (n_{splits} + 1)\ +\ n_{samples} \backslash \% (n_{splits} + 1)$ for the ith fold, while the test set will have the size of $n_{samples}\ /\ /\ (n_{splits} + 1)$.

However, you can change the train and test set size via the `max_train_size` and `test_size` arguments of the `TimeSeriesSplit` function. Additionally, there is also a `gap` argument that can be utilized to exclude G samples from the end of each train set, where G is the value needed to be specified by the developer.

You need to be aware that the Scikit-Learn implementation will always make sure that *there is no overlap between test sets*, which is actually not necessary. Currently, there is no way to enable the overlap between the test sets using the Scikit-Learn implementation. You need to *write the code from scratch* to perform that kind of strategy.

In this section, we learned about the unique characteristic of time-series data and how to perform a cross-validation strategy on it. There are other variations of the cross-validation strategy that haven't been covered in this book. If you are interested, you might find some pointers in the *Further reading* section.

Summary

In this chapter, we learned a lot of important things that we need to know regarding how to evaluate ML models properly. Starting from the concept of overfitting, numerous data splitting strategies, how to choose the best data splitting strategy based on the given situation, and how to implement each of them using the Scikit-Learn package. Understanding these concepts is important since you can't perform a good hyperparameter tuning process without applying the appropriate data splitting strategy.

In the next chapter, we will discuss hyperparameter tuning. We will not only discuss the definition but also several misconceptions and types of hyperparameter distributions.

Further reading

In this chapter, we have covered a lot of topics. However, there are still many uncovered interesting algorithms related to cross-validation due to the scope of this book. If you want to learn more about those algorithms and the implementation details of each of them, you can refer to this awesome page created by the Scikit-Learn authors at `https://scikit-learn.org/stable/modules/cross_validation.html`.

2

Introducing Hyperparameter Tuning

Every **machine learning** (**ML**) project should have a clear goal and success metrics. The success metrics can be in the form of business and/or technical metrics. Evaluating business metrics is hard, and often, they can only be evaluated after the ML model is in production. On the other hand, evaluating technical metrics is more straightforward and can be done during the development phase. We, as ML developers, want to achieve *the best technical metrics that we can get* since this is something that we can optimize.

In this chapter, we'll learn one out of several ways to optimize the chosen technical metrics, called **hyperparameter tuning**. We will start this chapter by understanding what hyperparameter tuning is, along with its goal. Then, we'll discuss the difference between a hyperparameter and a parameter. We'll also learn the concept of hyperparameter space and possible distributions of hyperparameter values that you may find in practice.

By the end of this chapter, you will understand the concept of hyperparameter tuning and hyperparameters themselves. Understanding these concepts is crucial for you to get a bigger picture of what will be discussed in the next chapters.

In this chapter, we'll be covering the following main topics:

- What is hyperparameter tuning?
- Demystifying hyperparameters versus parameters
- Understanding hyperparameter space and distributions

What is hyperparameter tuning?

Hyperparameter tuning is a process whereby we search for the best set of hyperparameters of an ML model from all of the candidate sets. It is the process of optimizing the technical metrics we care about. The goal of hyperparameter tuning is simply to get the *maximum evaluation score* on the validation set *without causing an overfitting issue*.

Hyperparameter tuning is one of the *model-centric* approaches to optimizing a model's performance. In practice, it is suggested to *prioritize data-centric* approaches over a model-centric approach when it comes to optimizing a model's performance. Data-centric means that we are focusing on cleaning, sampling, augmenting, or modifying the data, while model-centric means that we are focusing on the model and its configuration.

To understand why data-centric is prioritized over model-centric, let's say you are a cook in a restaurant. When it comes to cooking, no matter how expensive and fancy your kitchen setups are, if the ingredients are not in a good condition, it's impossible to serve high-quality food to your customers. In that analogy, ingredients refer to the data, and kitchen setups refer to the model and its configuration. No matter how fancy and complex our model is, if we do not have good data or features in the first place, then we can't achieve the maximum evaluation score. This is expressed in the famous saying, **garbage in, garbage out** (**GIGO**).

In model-centric approaches, hyperparameter tuning is performed after we have found the most suitable model framework or architecture. So, it can be said that *hyperparameter tuning is the ultimate step* in optimizing the model's performance.

Now that you are aware of hyperparameter tuning and its purpose, let's discuss hyperparameters themselves What actually is a hyperparameter? What is the difference between hyperparameters and parameters? We will discuss this in the next section.

Demystifying hyperparameters versus parameters

The *key difference* between a hyperparameter and a parameter is how its value is generated. A **parameter** value is generated by the model during the model-training phase. In other words, its value is *learned from the given data* instead of given by the developer. On the other hand, a **hyperparameter** value is *given by the developer* since it can't be estimated from the data.

Parameters are like the heart of the model. Poorly estimated parameters will result in a poorly performing model. In fact, when we said we are training a model, it actually means that we are providing the data to the model so that the model can estimate the value of its parameters, which is usually done by performing some kind of optimization algorithm. Here are several examples of parameters in ML:

- Coefficients (β_0, β_1, ... ,β_i, ... , β_n) in linear regression
- Weights (W^1, W^2, ..., W^i, ..., W^n) in a **multilayer perceptron** (**MLP**)

Hyperparameters, on the other hand, are a set of values that support the model-training process. They are defined by the developer without knowing the exact impact on the model's performance. That's why we need to perform hyperparameter tuning to get the best out of our model. The searching process can be done via exhaustive search, heuristic search, Bayesian optimization, or multi-fidelity optimization, which will be discussed in the following chapters. Here are several examples of hyperparameters:

- Dropout rate, number of epochs, and batch size in a **neural network** (**NN**)

- Maximum depth and splitting criterion in a decision tree

- Number of estimators in a random forest

You also need to be aware that there are models without hyperparameters or parameters, but not both of them. For instance, a linear regression model is a model that has only parameters but doesn't have any hyperparameters. On the other hand, **K-Nearest Neighbors** (**KNN**) is an instance of a model that doesn't contain any parameters but has a *k* hyperparameter.

More possible confusion may appear when we start writing our code and developing the ML model. In programming, arguments in a particular function or class are also often called parameters. What if we utilize a class that implements an ML model, such as a decision-tree model? What should we call the maximum depth or splitting criterion arguments that need to be passed to the class? Are they parameters or hyperparameters? The correct answer is *both*! They are *parameters to the class* and they are *hyperparameters to the decision-tree model*. It's just a matter of perspective!

In this section, we have learned what hyperparameters and parameters are, as well as what makes them different. In the next section, we will dive deeper into the realm of hyperparameters.

Understanding hyperparameter space and distributions

Hyperparameter space is defined as the universal set of possible hyperparameter value combinations—in other words, it is the space containing all possible hyperparameter values that will be used as the search space during the hyperparameter-tuning phase. That's why it is also often called the hyperparameter-tuning **search space**. This space is *predefined* before the hyperparameter-tuning phase so that the search will be performed only on this space.

For example, we want to perform hyperparameter tuning on a NN. Let's say we want to search what is the best value for the dropout rate, the number of epochs, and batch-size hyperparameters.

The dropout rate is bounded in nature. Its value can only be between 0 and 1, while for the number of epochs and batch-size hyperparameters, in theory, we can specify any positive integer value. However, there are other considerations that we need to think of. A higher batch size will usually produce a better model performance, but it will be bounded by the memory size of our physical computer. As for the number of epochs, if we go with too high a value, we will more likely be faced with an overfitting issue. That's why we need to *set boundaries* for the values of possible hyperparameters, which we call the hyperparameter space.

Hyperparameters can be in the form of *discrete or continuous* values. A discrete hyperparameter can be in the form of integer or string data types, while a continuous hyperparameter will always be in the form of real numbers or floating data types.

When defining a hyperparameter space, for *some hyperparameter-tuning methods*, it is not enough to only specify the possible values of each hyperparameter we care about. We also need to define what is the underlying **distribution** for each hyperparameter. Here, a distribution acts as some kind of *policy* that rules how likely it is that a specific value will be tested during the hyperparameter-tuning phase. If it is a uniform distribution, then all possible values have the same probability of being chosen.

There are many types of probability distributions that can be used: uniform, log-uniform, normal, log-normal, and many more. There is no best practice when it comes to choosing the appropriate distribution; you can just treat it as another hyperparameter. It is worth noting that there are distributions specifically for continuous hyperparameters, and there are also distributions for discrete ones. For discrete hyperparameter distribution, some distributions are specifically designed for discrete values—for instance, an integer uniform distribution—but there are also distributions that are adjusted from a continuous distribution. The latter types of discrete distributions usually have a *discretized* or *quantized* prefix on their name—for instance, a quantized uniform distribution.

It is also worth noting that *not all hyperparameters are equally significant* when it comes to impacting the model's performance—that's why it is recommended that you prioritize. We do not have to perform hyperparameter tuning on all of the hyperparameters of a model—just *focus on more important hyperparameters*.

In this section, we have learned about hyperparameter space and the concept of a hyperparameter distribution and looked at examples of hyperparameter distributions you may find in practice.

Summary

In this chapter, we have learned all we need to know about hyperparameter tuning, starting from what it is, what is its goal, and when we should perform hyperparameter tuning. We have also discussed the difference between hyperparameters and parameters, the concept of hyperparameter space, and the concept of hyperparameter distributions. Having a clear picture of the concept of hyperparameter tuning and hyperparameters themselves will help you a lot in the following chapters.

As stated previously, we will discuss all of the four categories of hyperparameter-tuning methods in this book. In *Chapter 3*, *Exploring Exhaustive Search*, we will start discussing the first group and the most widely used hyperparameter-tuning methods in practice. There will be both high-level and detailed explanations to help you understand each of the methods more easily.

3
Exploring Exhaustive Search

Hyperparameter tuning doesn't always correspond to fancy and complex search algorithms. In fact, a simple `for` loop or manual search based on the developer's instinct can also be utilized to achieve the goal of hyperparameter tuning, which is to get the maximum evaluation score on the validation score without causing an overfitting issue.

In this chapter, we'll discuss the first out of four groups of hyperparameter tuning, called an **exhaustive search**. This is the *most widely used* and *most straightforward* hyperparameter-tuning group in practice. As explained by its name, hyperparameter-tuning methods that belong to this group work by *exhaustively searching* through the hyperparameter space. Except for one method, all of the methods in this group are categorized as **uninformed search** algorithms, meaning they are not learning from previous iterations to have a better search space in the future. Three methods will be discussed in this chapter: manual search, grid search, and random search.

By the end of this chapter, you will understand the concepts of each of the hyperparameter-tuning methods that belong to the exhaustive search group. You will be able to explain these methods with confidence when someone asks you about them, in both a high-level and detailed fashion, along with the pros and cons. More importantly, you will be able to apply all of the methods with high confidence in practice. You will also be able to understand what's happening if there are errors or unexpected results and understand how to set up the method configuration to match your specific problem.

The following main topics will be covered in this chapter:

- Understanding manual search
- Understanding grid search
- Understanding random search

Understanding manual search

Manual search is the most straightforward hyperparameter-tuning method that belongs to the exhaustive search group. In fact, it's not even an algorithm! There's no clear rule on how to perform this method. As its name would suggest, a manual search is performed based on your instinct. You simply have to tweak the hyperparameters until you are satisfied enough with the result.

This method is *the one exception* mentioned before in the introduction of this chapter. Except for this method, other methods in the exhaustive search group are categorized as **uninformed search methods**. You may already know the reason why this method is the exception. It's because the developer themselves learn what is the impact of changing a particular or a set of hyperparameters in each iteration. In other words, they learn from previous iterations to have a (hopefully) better "hyperparameter space" in the next iterations.

To perform a manual search, do the following:

1. Split the original full data into train and test sets (see *Chapter 1, Evaluating Machine Learning Models*).

2. Specify initial hyperparameter values.

3. Perform cross-validation on the train set (see *Chapter 1, Evaluating Machine Learning Models*).

4. Get the cross-evaluation score.

5. Specify new hyperparameter values.

6. Repeat *steps 3-5* until you are satisfied enough.

7. Train on the full training set using the final hyperparameter values.

8. Evaluate the final trained model on the test set.

Although this method seems very straightforward and easy to do, it is actually *the other way around for a beginner*. This is because you need to really understand how the model works and the usage of each hyperparameter. It is also worth noting that, when it comes to manual search, there is *no clear definition of the hyperparameter space*. The hyperparameter space can be surprisingly narrow or vast, based on the developer's willingness and initiative to experiment with it.

Here is a list of pros and cons of the manual search hyperparameter-tuning method:

Pros	Cons
For a skilled developer, combining this method with other methods can help to reduce the experiment time.	Hard to guess a good hyperparameter value, even for someone who really understands how the model works.
	Performing a manual search alone is time-consuming.

Figure 3.1 – Manual search: pros and cons

Now that you are aware of how manual search works, along with the pros and cons, we will learn the simplest automated hyperparameter-tuning strategy, which will be discussed in the next section.

Understanding grid search

Grid search is the *simplest automated hyperparameter-tuning method* that ever existed. Apart from the fancy name, grid search is basically just a *nested* `for` loop that tests all possible hyperparameter values in the search space. Although many packages have grid search as one of their hyperparameter-tuning method implementations, it is super easy to write your own code from scratch to implement this method. The name *grid* comes from the fact that we have to test the whole hyperparameter space just like creating a grid, as illustrated in the following diagram.

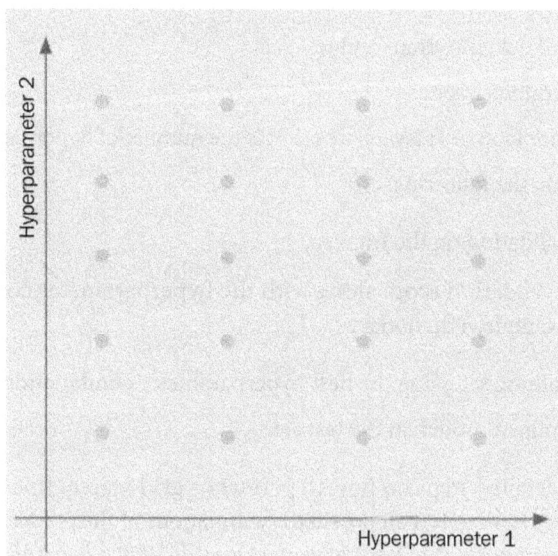

Figure 3.2 – Grid search illustration

For example, let's say we want to perform hyperparameter tuning using the grid search method on a random forest. We decide to focus only on the number of estimators, splitting criterion, and maximum tree-depth hyperparameters. Then, we can specify a list of possible values for each of the hyperparameters. Let's say we define the hyperparameter space as follows:

- Number of estimators: `n_estimators = [25, 50, 100, 150, 200]`
- Splitting criterion: `criterion = ["gini", "entropy"]`
- Maximum depth: `max_depth = [3, 5, 10, 15, 20, None]`

Notice that for the grid search method, we do not have to specify the underlying distribution of the hyperparameters. We simply create a list of all values that we want to test on for each hyperparameter. Then, we can call the grid search implementation from our favorite package or write the code for grid search by ourselves, as illustrated in the following snippet:

```
for n_est in n_estimators:
        for crit in criterion:
        for m_depth in max_depth:
        #perform cross-validation here
```

In this example, we create a nested `for` loop consisting of three layers, each for the hyperparameter in our search space. To perform a grid search in general, do the following:

1. Split the original full data into train and test sets.
2. Define the hyperparameter space.
3. Construct a nested for loop of H layers, where H is the number of hyperparameters in the space.
4. Within each loop, do the following:

    *   Perform cross-validation on the train set
    *   Store the cross-validation score along with the hyperparameter combination in a data structure—for example, a dictionary

5. Train on the full training set using the best hyperparameter combination.
6. Evaluate the final trained model on the test set.

As you can see from the detailed steps on how to perform a grid search, this method is actually a *brute-force* method since we have to test all possible combinations of the predefined hyperparameter space. That's why it is very important to have a proper or *well-defined hyperparameter space*. If not, then we will waste a lot of time testing all of the combinations.

Here is a list of pros and cons of the grid search hyperparameter-tuning method:

Pros	Cons
Very easy to implement from scratch	**Curse of dimensionality (COD)**
Allowed to test all possible combinations	Possible to miss better hyperparameter combinations outside the predefined search space

Figure 3.3 – Grid search: pros and cons

The *COD* in *Table 3.2* means that adding another value to the hyperparameter space will *exponentially increase* the experiment time. Let's use the preceding example where we performed hyperparameter tuning on a random forest. In our initial hyperparameter space, there are $5 \cdot 2 \cdot 6 = 60$ combinations we have to test. If we add just another value to our space—let's say we add 30 to the max_depth list—there will be $5 \cdot 2 \cdot 7 = 70$ combinations or an additional 10 combinations that we have to test. This exponential behavior will even become more apparent when we have a bigger hyperparameter space! Sadly, it is also possible that after defining a big hyperparameter space and spending a long time performing hyperparameter tuning, we can still *miss better hyperparameter values* since they are located outside of the predefined space!

In this section, we have learned what grid search is, how it works, and what the pros and cons are. In the next section, we will discuss the last hyperparameter-tuning method that is categorized in the exhaustive search group: the random search method.

Understanding random search

Random search is the third and the last hyperparameter-tuning method that belongs in the exhaustive search group. It is a *simple method but works surprisingly well* in practice. As implied by its name, random search works by *randomly selecting hyperparameter values* in each iteration. There's nothing more to it. The selected set of hyperparameters in the previous iteration will not impact how the method selects another set of hyperparameters in the following iterations. That's why random search is also categorized as an *uninformed search* method.

You can see an illustration of the random search method in the following diagram:

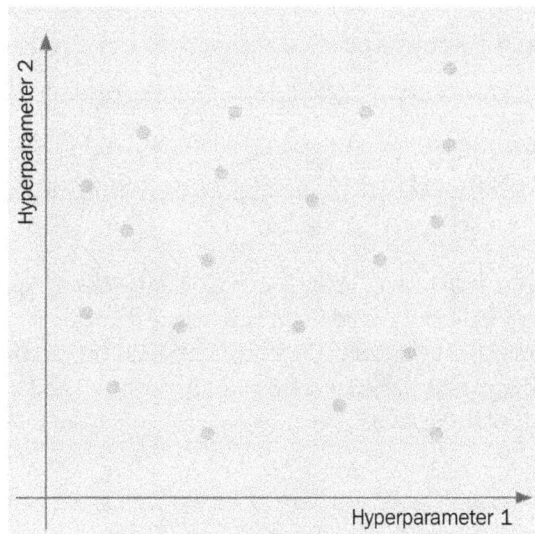

Figure 3.4 – Random search illustration

Random search usually works better than grid search when we have little or no idea of the proper hyperparameter space for our case, and this applies most of the time. Compared to grid search, random search is also more efficient in terms of computing cost and in finding the optimal set of hyperparameters. This is because we do not have to test each of the hyperparameter combinations; we can just let it run stochastically—or, in layman's terms, we can just *let luck play its part*.

You may wonder how picking a random set of hyperparameters can lead to a better tuning result compared to grid search most of the time. It is actually not the case if the predefined hyperparameter space is exactly the same as the one we provide to the grid search method. We have to provide a *bigger hyperparameter space* in order to support random search to play its role. A bigger search space doesn't always mean we have to *increase the dimensionality*, either by widening existing hyperparameters' range or adding new hyperparameters. We can also create a bigger hyperparameter space by *adding granularity* to it.

It is also worth noting that, unlike grid search, which doesn't require defining the hyperparameter's distribution when defining a search space, in random search, it is suggested to *define the distribution of each hyperparameter*. In some package implementations, if you do not specify the distribution, it will default to the uniform distribution. We will discuss more on the implementation part from *Chapter 7, Hyperparameter Tuning via Scikit* to *Chapter 10, Advanced Hyperparameter Tuning with DEAP and Microsoft NNI*

Let's use a similar example to what we saw in the *Understanding grid search* section to get a better understanding of how random search works. Apart from focusing on the number of estimators, splitting criterion, and maximum tree depth, we will also add a minimum samples split hyperparameter to our space. Unlike grid search, we have to also provide a distribution of each of the hyperparameters when defining a search space. Let's say we define the hyperparameter space as follows:

- Number of estimators: `n_estimators = randint(25,200)`
- Splitting criterion: `criterion = ["gini", "entropy"]`
- Maximum depth: `max_depth = [3, 5, 10, 15, 20, None]`
- Minimum samples split: `min_samples_split = truncnorm(a=1, b=5, loc=2, scale=0.5)`

As you can see, compared to the search space in the *Understanding grid search* section, we are *increasing the size of the space* by adding granularity and adding a new hyperparameter. We add granularity for the `n_estimators` hyperparameter by utilizing the `randint` uniform random integer distribution, ranging from 25 to 200. This means we can test any value between 25 and 200, where all of them will have the same probability of being tested.

Apart from increasing the size of the search space by adding granularity, we also add a new hyperparameter called `min_samples_split`. This hyperparameter has the `truncnorm` distribution or **truncated normal distribution**, which basically—as implied by its name—is a modified normal distribution bounded on a particular range. In this case, the range is bounded on a range from `a=1` and `b=5`, with a mean of `loc=2` and standard deviation of `scale=0.5`.

As for `criterion` and `max_depth`, we are still using the same configuration as the previous search space. Note that *not specifying any distribution* means we are applying uniform distribution to the hyperparameter, where all values will have the same probability of being tested. For now, you don't have to worry about what are the available distributions and how to implement them, since we will also discuss them from *Chapter 7, Hyperparameter Tuning via Scikit* to *Chapter 10, Advanced Hyperparameter Tuning with DEAP and Microsoft NNI*.

In random search, apart from the need to define a hyperparameter space, we also need to define a hyperparameter for this method itself, which is called the **number of trials**. This hyperparameter will *control how many trials or iterations* we want to perform on the predefined search space. This hyperparameter is needed since we *are not aiming to test all possible combinations* in the space; if we were, then it would be the same grid search method. It is also worth noting that since this method has a stochastic nature, we also need to specify a *random seed* to get the exact same result every time we run the code.

Unlike grid search, it is quite cumbersome to implement this method from scratch, although it is possible to do so. Therefore, many packages support the implementation of the random search method. Regardless of the implementation variations, in general, random search works like this:

1. Split the original full data into train and test sets.

2. Define the number of trials and a random seed.

3. Define a hyperparameter space with the accompanied distributions.

4. Generate an iterator consisting of random hyperparameter combinations with the number of elements equal to the defined number of trials in Step 2.

5. Loop through the iterator, where the following actions will be performed within each loop:

   - Getting the hyperparameter combination for this trial from the iterator

   - Performing cross-validation on the train set

   - Storing the cross-validation score along with the hyperparameter combination in a data structure—for example, a dictionary

6. Train on the full training set using the best hyperparameter combination.

7. Evaluate the final trained model on the test set.

Please note that it is guaranteed there is *no duplicate* in the generated hyperparameter combinations in *Step 4*.

Here is a list of pros and cons of the random search hyperparameter-tuning method:

Pros	Cons
Compared to grid search, this method is more efficient in terms of computing cost and getting optimal hyperparameters.	Results in high variance during the process.
Works well to discover unexpected optimal hyperparameter combinations.	Sometimes, more time is needed to get the optimal hyperparameters.

Figure 3.5 – Random search: pros and cons

The random search produces high variance during the process due to the property of *uninformed search* methods. There is no way for the random search to learn from past experiences so that it can learn better and be more effective in the next iterations. In *Chapter 6, Exploring Multi-Fidelity Optimization*, we will learn other *variations of grid search and random search* that are categorized as informed search methods.

In this section, we have learned all you need to know about random search, starting from what it is, how it works, what makes it different from grid search, and the pros and cons of this method.

Summary

In this chapter, we have discussed the first out of four groups of hyperparameter-tuning methods, called the exhaustive search group. We have discussed manual search, grid search, and random search. We not only discussed the definition of those methods, but also how those methods work at both a high level and a technical level, and what are the pros and cons for each of them. From now on, you should be able to explain these exhaustive search methods with confidence when someone asks you about them and apply all of the exhaustive search methods with high confidence in practice.

In the next chapter, we will start discussing Bayesian optimization, the second group of hyperparameter-tuning methods. The goal of the next chapter is similar to this chapter, which is to give a better understanding of methods belonging to the Bayesian optimization group so that you can utilize those methods with high confidence in practice.

4

Exploring Bayesian Optimization

Bayesian optimization (BO) is the second out of four groups of hyperparameter tuning methods. Unlike grid search and random search, which are categorized as uninformed search methods, all of the methods that belong to the BO group are categorized as **informed search** methods, meaning they are learning from previous iterations to (hopefully) provide a better search space in the future.

In this chapter, we will discuss several methods that belong to the BO group, including **Gaussian process (GP)**, **sequential model-based algorithm configuration (SMAC)**, **Tree-structured Parzen Estimators (TPE)**, and Metis. Similar to *Chapter 3, Exploring Exhaustive Search*, we will discuss the definition of each method, the differences between them, how they work, and the pros and cons of each method.

By the end of this chapter, you will be able to explain BO and its variations when someone asks you. You will not only be able to explain what they are, but also how they work, in a high-level and technical way. You will also be able to tell the differences between them, along with the pros and cons of each of the methods. Furthermore, you will experience a crucial benefit once you understand the ins and outs of each method; that is, you will be able to understand what's happening if there are errors or unexpected results and understand how to set up the method configuration to match your specific problem.

In this chapter, we will cover the following main topics:

- Introducing BO
- Understanding BO GP
- Understanding SMAC
- Understanding TPE
- Understanding Metis

Introducing BO

BO is categorized as an informed search hyperparameter tuning method, meaning the search is learning from previous iterations to have a (hopefully) better subspace in the next iterations. It is also categorized as the **sequential model-based optimization (SMBO)** group. All SMBO methods work by sequentially updating probability models to estimate the effect of a set of hyperparameters on their performance based on historical observed data, as well as suggesting new hyperparameters to be tested in the following trials.

BO is a popular hyperparameter tuning method due to its *data-efficient* property, meaning it needs a relatively small number of samples to get to the optimal solution. You may be wondering, how exactly does BO get this ground-breaking data-efficient property? This property exists thanks to BO's ability to learn from previous iterations. BO can learn and predict which subspace is worth visiting in the future by utilizing a **probabilistic regression model**, which acts as the *cheap cloned version of the expensive objective function*, and an **acquisition function**, which governs which *set of hyperparameters should be tested* in the next iteration.

The objective function is just a function that takes hyperparameter values as input and returns the cross-validation score (see *Chapter 1, Evaluating Machine Learning Models*). We do not know what the output of the objective function for all possible hyperparameter values is. If we did, there would be no need to perform hyperparameter tuning. We could just use that function to get the hyperparameter values, which results in the highest cross-validation score. That's why we need a probabilistic regression model, to approximate the objective function by fitting a set of known hyperparameter and cross-validation score value pairs (see *Figure 4.1*). The approximation concept is *similar to the concept of ML-based regressor* models, such as random forest, linear regression, and many more. First, we fit the regressor to the samples of independent and dependent variables; then, the model will try to *learn* from the data, which in the end can be used to predict new given data. The probabilistic regression model is also often called the **surrogate model**:

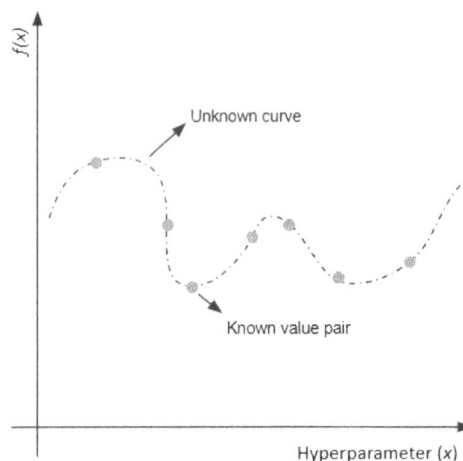

Figure 4.1 – Illustration of the probabilistic regression model, M

The acquisition function *governs which subspace we should search in the next iteration*. Thanks to this function, BO enables us to learn from past experiences and have fewer hyperparameter tuning iterations compared to random search, in general.

> **Important Note**
>
> Remember that, to get the cross-validation score, we need to perform multiple training and evaluation processes (see *Chapter 1, Evaluating Machine Learning Models*). This is an *expensive process* when you have a big, complex model with a large amount of training data. That's why the acquisition function plays a big role here.

In general, BO works as follows:

1. Split the original full data into train and test sets. (See *Chapter 1, Evaluating Machine Learning Models*).

2. Define the hyperparameter space, H, with the accompanied distributions.

3. Define the objective function, f, based on the train set.

4. Define the stopping criterion. Usually, the number of trials is used. However, it is also possible to use the time taken or convergence as the stopping criterion.

5. Initializes the empty set, D, which will be used to store the initial pairs of hyperparameter values and cross-validation scores, as well as the resulting pairs suggested by the acquisition function, A.

6. Initialize several pairs of hyperparameter values and cross-validation scores and store them in D.

7. Fit the probabilistic regression model/surrogate model, M, using the value pairs in D.

8. Sample the next set of hyperparameters by utilizing the acquisition function, A:

    I. Perform optimization on the acquisition function, A, with the help of the surrogate model, M, to sample which hyperparameters are to be passed to the acquisition function.

    II. Get the expected optimal set of hyperparameters based on the acquisition function, A.

9. Compute the cross-validation score using the objective function, f, based on the output from *Step 8*.

10. Add the hyperparameters and cross-validation score pair from *Step 8* and *Step 9* to set D.

11. Repeat *Steps 7* to *10* until the stopping criterion is met.

12. Trains on the full training set using the final hyperparameter values.

13. Evaluate the final trained model on the test set.

You can initialize the hyperparameter values and cross-validation scores, as shown in *Step 6*, using several sampling strategies. The most straightforward and go-to way, in practice, is to just perform **random sampling**. However, there are also other methods that you may consider during your experiments, such as the **quasi-random** or **Latin hypercube** sampling methods.

Similar to random search, we also need to define the distribution of each hyperparameter in BO. You may wonder if BO can also work on a non-numerical type of hyperparameter. The answer is *based on the probabilistic regression model* you are using. There are several surrogate models you can choose from. Those options will be discussed in the next three sections of this chapter, and they include **GP**, **Tree-structured Parzen Estimator** (**TPE**), random forest, extra trees, or other ML-based regressors. In this book, we will discuss the random forest regressor that's implemented in the SMAC model.

It is also worth noting that the optimization process in *Step 8* can be *replaced with a random search*. So, instead of performing some kind of second-order optimization method, we can randomly sample sets of hyperparameters from the search space and pass them onto the acquisition function. Then, we can get the optimal set of hyperparameters based on the output from the acquisition function. When using random search in this step, we still utilize the acquisition function to govern which subspace we should search for in the next iteration, but we add some random behavior to it, with the hope that we can escape the local optimum and converge toward the global optimum.

The first and the most popular acquisition function is **expected improvement** (**EI**), which is defined as follows:

$$EI(x) = \big(\mu(x) - f(x^\cdot)\big)\Phi(Z) + \sigma(x)\phi(Z) \quad \text{when} \ \ \sigma(x) \neq 0$$

$$EI(x) = 0 \quad \text{when} \ \ \sigma(x) = 0$$

Here, $Z = \frac{\mu(x) - f(x^\hat{})}{\sigma(x)}$, $\Phi(Z)$ and $\phi(Z)$ are the cumulative distribution and probability density functions of the standard normal distribution, respectively. $\mu(x)$ and $\sigma(x)$ represent the expected performance and the uncertainty, respectively, that are captured by the surrogate model. Finally, $f(x^\cdot)$ represents the current best value of the objective function.

Implicitly, the EI acquisition function enables BO methods to have the *exploration versus exploitation trade-off property*. This property can be achieved by two terms competing within the formula. When the value of the first term is high, meaning the expected performance, $\mu(x)$, is higher than the current best value, $f(x^\cdot)$, EI will favor the exploitation process. On the other hand, when the uncertainty is very high, meaning we have a high value of $\sigma(x)$, EI will favor the exploration process. By exploitation, this means that the acquisition function will recommend the set of hyperparameters that possibly get a higher value of the objective function, f. In terms of exploration, this means that the acquisition function will recommend the set of hyperparameters from the subspace that we haven't explored yet.

You can imagine this exploration and exploitation trade-off as when you are craving some food. Let's say you want to have lunch with your brother today. Imagine the following two scenarios:

- "Hey bro, let's have lunch at our favorite restaurant today!"

- "Hey bro, have you heard of the new restaurant up there? Why don't we try it for lunch?"

In the first scenario, you choose to eat at your favorite restaurant since you are confident that there is nothing wrong with the food and, more importantly, you are *confident about the taste of the food and the overall experience* of eating at that restaurant. This first scenario best explains what we call the exploitation process. In the second scenario, you *don't have any idea what the overall experience* of eating at that new restaurant is. It may be worse than your favorite restaurant, but it may also potentially be your new favorite restaurant! This is what we call the exploration process.

> **Important Note**
>
> In some implementations, such as in the **Scikit-optimize** package, there is a hyperparameter that enables us to *control how much we are leaning toward exploitation* compared to exploration. In Scikit-optimize, the sign of the EI function is negative. This is because the package *treats the optimization problem as the minimization problem* by default.

In our previous explanation, we treated the optimization problem as the maximization problem since we wanted to get the highest cross-validation score possible. Don't confuse this with the minimization versus maximization problem – just choose what best describes the problem you will be facing in practice!

The following is the EI acquisition function that's implemented in the Scikit-optimize package:

$$-EI(x) = (f(x\hat{}) - \mu(x) - \delta)\Phi(Z) + \sigma(x)\phi(Z)$$

As you can see in the first term, the value of δ will control how big our tendency is toward exploitation compared to exploration. The smaller the δ value is, the more we lean toward exploitation. We will learn more about the implementation part of BO using Scikit or other packages from *Chapter 7, Hyperparameter Tuning via Scikit* to *Chapter 10, Advanced Hyperparameter Tuning with DEAP and Microsoft NNI*.

To get a better understanding of how the exploration and exploitation trade-off happens during the hyperparameter tuning phase, let's look at an example. Let's say, for instance, we are using the GP surrogate model to estimate the following objective function. There's no need to worry about what and how GP works for now; we will discuss it in more detail in the next section:

$$f(x) = \cos(6x) \cdot (1 - \sin(x^5)) + \varepsilon$$

Here, ε is a noise that follows the standard normal distribution. The following is a plot of this function within the range of $[-2,2]$. Note that, in this example, we are assuming that we know what the true objective function is. However, in practice, this function is unknown:

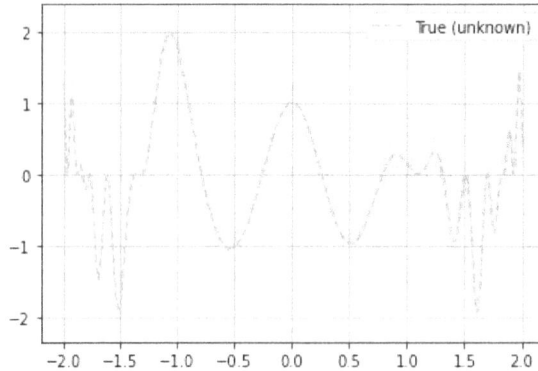

Figure 4.2 – Plot of the objective function, f(x)

Let's say we are using the EI as the acquisition function, setting the number of trials as 15, setting the initial number of points as 5, and setting the δ value to 0.01. You can see how the fitting process works for the first five trials in the following figure:

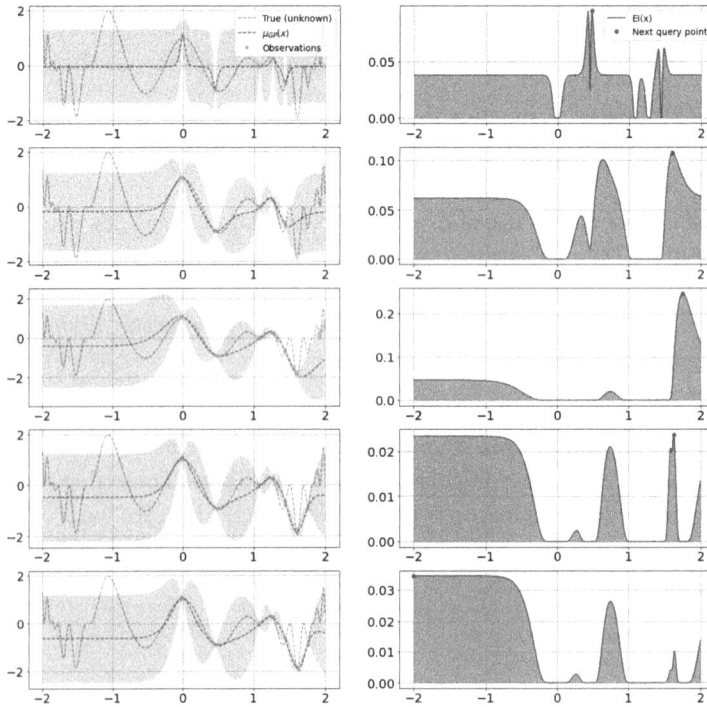

Figure 4.3 – GP and EI illustration, δ = 0.01

Each row in the preceding figure corresponds to the first until the fifth trial. The left column contains information on the objective function (*red dashed line*), the GP surrogate model approximation of the objective function (*green dashed line*), how sure the approximation is (*green transparent area*), and the observed points up to each trial (*red dots*). The right column contains information on the EI acquisition function values (*blue line*) and the next point (*blue dot*) to be included in the next trials.

Let's run through each of the rows in *Figure 4.3* so that you understand how it works. In the first trial (*see the first row from the top in the left column*), we initialize five random sample points – or hyperparameter values, in the context of hyperparameter tuning – and fit the GP model based on those five points. Remember that the GP model doesn't know the actual objective function; the only information it has is just those five random points. Then (*see the first row from the top in the right column*), based on the fitted GP model, we get the value of the EI acquisition function across the space. In this case, the space is just a range – that is, $[-2,2]$. We also get the point to be included in the next trials, which in this case is around point 0.5.

In the second trial, we utilize the point suggested by the EI acquisition function and fit the GP model again based on the six sample points we have (*see the second row from the top in the left column*). If you compare the GP approximation of the second trial with the first trial, you will see that it is closer to the true objective function. Next (*see the second row from the top in the right column*), we repeat the same process, which is to generate the EI function value across the space and the point to be included in the next trial. The suggested point in this step is around 0.7.

We keep repeating the same process until the stopping criteria are met, which in this case is 15 trials. The following plot shows the result after 15 trials. It is much better than the approximation in the first trial (*see the green dashed line*)! You can also see that there are some ranges of x where the confidence of the GP approximation is high, such as around points -1.5 and 1.6:

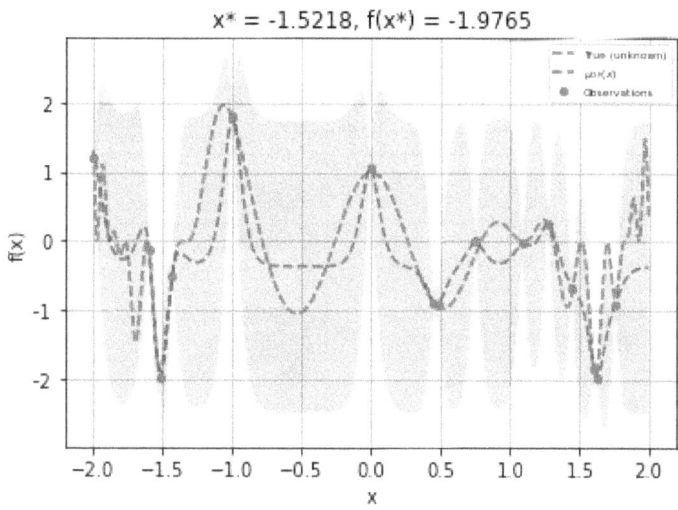

Figure 4.4 – Result after 15 trials, δ = 0.01

Based on the preceding plot, the final suggested point, or the hyperparameter value, is -1.5218, which results in the value of the objective function being equal to -1.9765. Let's also look at the convergence plot from the first until the last trial. From the following convergence plot, we can see how our surrogate model and acquisition function help us get the minimum value of the objective function based on all the trials:

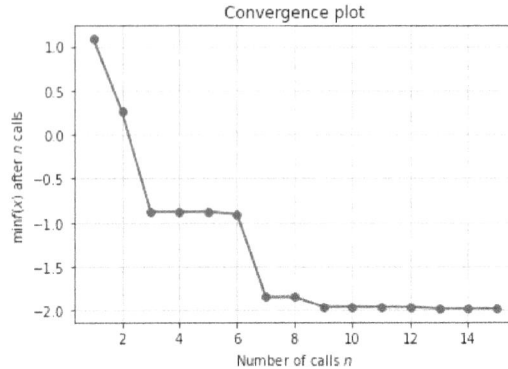

Figure 4.5 – Convergence plot

Now, let's try to change the value of δ to a lower value than what we had previously to see how the EI acquisition function will favor exploitation more than exploration. Let's set the δ value to be 1,000 times lower than the previous value. Note that we only change the δ value and leave the other setups as-is:

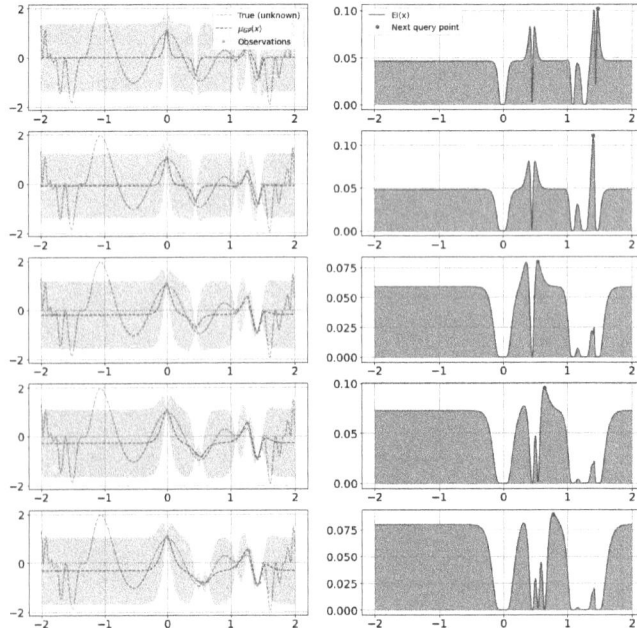

Figure 4.6 – GP and EI illustration, $\delta = 0.00001$

As you can see, the EI acquisition function suggested most of the points in a range between **0.5** and **1.4**. The acquisition function doesn't suggest exploring the [1.5,1.6] range, although we can get a much lower objective function value in that range. This happens because there are no initial random points in that range, and we favor exploitation a lot in this example. The following plot shows the final results after 15 trials. In this case, we get a worse result when we favor more exploitation over exploration. However, this is not always the case. *You have to experiment* since different data, different objective functions, a different hyperparameter space, and different implementations may result in different conclusions:

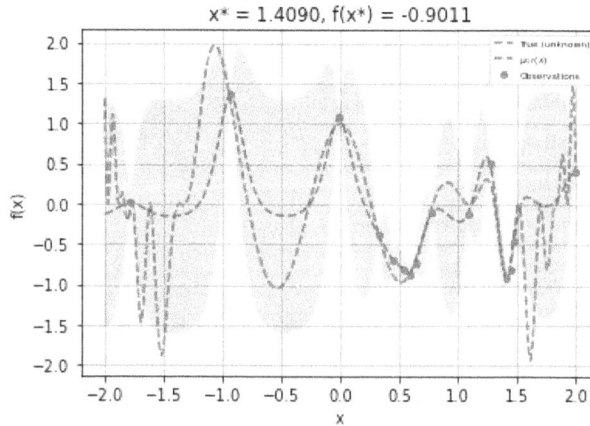

Figure 4.7 – Result after 15 trials, δ = 0.00001

Now, let's see what the impact is if we set the δ value to 100, which in this case means that we favor exploration more than exploitation. Similar to the previous trial, after running 15 trials, we got the following results:

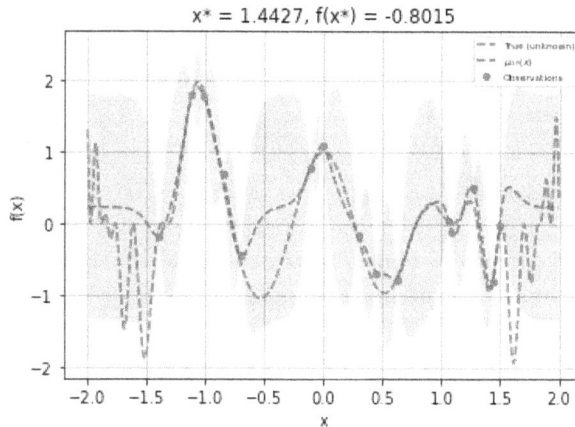

Figure 4.8 – Result after 15 trials, δ = 100

As you can see, the points that are suggested by the acquisition function (*the red dots*) are all over the place. This is because we set such a high δ value. This means that the acquisition function's outputs will suggest points in the space that haven't been observed yet. We will learn how to produce the plots shown here in *Chapter 7, Hyperparameter Tuning via Scikit*.

Besides the EI acquisition function, there are also other popular acquisition functions that you may consider using, including **Probability of Improvement (PI)** and **Upper Confidence Bound (UCB)**.

PI is the acquisition function that existed before EI. It is simpler than EI – in fact, the formula of $PI(x)$ is derived based on the following simple definition of *improvement*:

$$I(x) = \max\big(0, \mu(x) - f(x^\wedge)\big)$$

The idea of $I(x)$ is to return the size of improvement, if there is improvement between the expected performance and the current best performance, or just return zero if there is no improvement. Based on $I(x)$, we can define PI as follows:

$$PI(x) = \Phi(Z) = \Phi\left(\frac{\mu(x) - f(x^\wedge)}{\sigma(x)}\right) \text{ when } \sigma(x) \neq 0$$

$$PI(x) = 0 \text{ when } \sigma(x) = 0$$

The problem with PI is that it will *give the same reward for all sets of hyperparameters*, so long as there's an improvement compared to the current best value, $f(x^\wedge)$, no matter how big the improvement is. This behavior is not very preferable in practice since it can *guide us to the local minima and get us stuck in there*. If you are familiar with calculus and statistics, you will realize that EI is just the expectation over $I(x)$, as shown here:

$$EI(x) = \mathbb{E}[I(x)] = \int_{-\infty}^{\infty} I(x)\phi(z)dz$$

Here, $\phi(z)$ is the probability density function of the standard normal distribution. Unlike PI, the *EI acquisition function will take the size of improvement into account*.

As for the UCB, it is very straightforward compared to others. We have the power to control the trade-off between exploration and exploitation by ourselves via the λ parameter. This acquisition function can be defined as follows:

$$UCB(x; \lambda) = \mu(x) + \lambda \cdot \sigma(x)$$

As you can see, UCB *doesn't take into account the current best value* of the objective function. It only considers the expected performance and the uncertainty captured by the surrogate model. You can control the exploration and exploitation trade-off by changing the λ value. If you want to lean toward exploring the search space, then you can increase the value of λ. However, if you want to focus more on the set of hyperparameters that are expected to perform well, then you can decrease the value of λ.

Apart from the variations of surrogate model and acquisition functions, there are also other variations of BO methods based on modifying the algorithm itself, including Metis and **Bayesian optimization and HyperBand (BOHB)**. We will discuss Metis in the *Understanding Metis* section and BOHB in *Chapter 6, Exploring Multi-Fidelity Optimization*.

The following are the pros and cons of BO hyperparameter tuning, in general, compared to other hyperparameter tuning methods:

Pros	Cons
Able to handle expensive objective functions, especially when they have good initial points.	Hard to set up; a good understanding of the method is needed to get decent results.
Data-efficient; takes fewer iterations than random search when it has good initial points.	Tends to have a slower warm-up process compared to random search for a huge hyperparameter space.
Has the exploration versus exploitation trade-off feature.	
Fully automated; guidance from humans is not needed during the search process.	

Figure 4.9 – Pros and cons of BO

BO can handle expensive objective functions and is more data-efficient and arguably better than random search when it has good initial points. You can utilize the set of hyperparameters we used for the initial points up to *Step 6* from the procedure mentioned at the beginning of this section. However, if you don't have that privileged access, BO still can outperform random search if you give the method some more time since it has to build a good surrogate model first from scratch, especially if you have a huge hyperparameter space. Once BO has built a good surrogate model, it tends to work faster than random search to find the optimal set of hyperparameters.

There is also another way to speed up the relatively slow warm-up process of BO. The idea is to adopt a **meta-learning** procedure to initialize the initial set of hyperparameters by learning from meta-features in other, similar datasets.

> Speeding Up BO's Warm-Up
>
> See the following paper for more information: *Efficient and Robust Automated Machine Learning*, by Matthias Feurer, Aaron Klein, Katharina Eggensperger, Jost Springenberg, Manuel Blum, Frank Hutter (https://papers.nips.cc/paper/2015/hash/11d0e62872 02fced83f79975ec59a3a6-Abstract.html).

BO also has a nice feature that random search doesn't have – the ability to control the exploration and exploitation trade-off, as explained previously in this section. This feature enables BO to do more than just constantly explore, as random search does.

Now that you are aware of what BO is, how it works, what its important components are, and the pros and cons of this method, we will dive deeper into the variations of BO in the following sections.

Understanding BO GP

Bayesian optimization Gaussian process (BOGP) is one of the variants of the BO hyperparameter tuning method. It is well-known for its good capability in describing the objective function. This variant is very popular due to the unique *analytically tractable* nature of the surrogate model and its ability to produce relatively accurate approximation, even with only a few observed points.

However, BOGP has limitations. It *only works on continuous hyperparameters*, not on the discrete or categorical types of hyperparameters. It is not recommended to use BOGP when you need a lot of iterations to get the optimal set of hyperparameters, especially when you have a large number of samples. This is BOGP has a $O(N^3)$ runtime, where N is the number of samples. If you have *more than 10 hyperparameters* to be optimized, the common belief is that BOGP is not the right hyperparameter tuning method for you.

Having GP as the surrogate model means that we utilize GP as the *prior* for our objective function. Then, we can utilize the prior along with a *likelihood model* to compute the *posterior* that we care about. All of these nerdy terms can easily be understood if we are familiar with the famous **Bayes Theorem**.

Bayes Theorem allows us to calculate the probability of an event, given a specific condition, by utilizing our previous knowledge or common belief that we have. Formally, Bayes Theorem is defined as follows:

$$P(\Theta|data) = \frac{P(data|\Theta) \cdot P(\Theta)}{P(data)}$$

Here, Θ is the event we want to know the probability of, and $data$ refers to the specific condition we mentioned previously. The left-hand side of the equation, $P(\Theta|data)$, is what we called as the posterior. $P(\Theta)$ is the prior and $P(data|\Theta)$ is what we call the likelihood model. Finally, $P(data)$ is just a constant to ensure that the resulting value of this formula is bound in the range of $[0,1]$.

To understand Bayes Theorem, let's walk through an example. Let's say we want to know the probability of you eating at your favorite restaurant, given that today's weather is sunny. In this example, you eating at your favorite restaurant is the event we are interested in. This is Θ in the equation. The information that today is sunny refers to $data$ in the equation.

Let's say you are eating at your favorite restaurant for 40 out of 100 days. This means that before knowing what today's weather is, your $P(\Theta)$ is equal to $\frac{40}{100} = 0.4$. Let's also assume that out of 100 days, there are 30 sunny days. Then, the $P(data)$ value is equal to $\frac{30}{100} = 0.3$. Based on your experience of eating at your favorite restaurant, you have realized that you ate in the sunny weather condition 20 out of 40 times. Thus, the likelihood, $P(data|\Theta)$, is equal to $\frac{20}{40} = 0.5$. Using all of this information, we can calculate the probability of you eating at your restaurant, given that today's weather is sunny, as

$$P(\Theta|data) = \frac{0.5 \cdot 0.4}{0.3} = \frac{2}{3}$$

Now, we are ready to revisit the GP. BOGP utilizes GP as the surrogate model. GP as the surrogate model means that we utilize it as the prior of our objective function, which implies that the *posterior distribution is also a GP*. You can think of GP as a generalization of a Gaussian distribution that you are familiar with. Unlike Gaussian distribution, which describes the distribution of a random variable, *GP describes the distribution over functions*. Similar to the Gaussian distribution that is accompanied by the mean and variance of the random variable, GP is also accompanied by the *mean and covariance* of the function. As for the *likelihood*, we assume that the objective function, *f*, follows a normal likelihood with noise:

$$y = f(x) + \varepsilon$$

$$\varepsilon \sim N(0, \sigma_\varepsilon^2)$$

Then, we can describe $f_{1:n} = \{f(x_1), f(x_2), ..., f(x_n)\}$, or the values of our objective function for all n samples. as a GP with a mean function of $m(x_{1:n})$ and a covariance kernel, K, sized $n \times n$, which is defined as follows:

$$f_{1:n} \sim N(m(x_{1:n}), K)$$

The distribution of prediction from GP also follows the Gaussian distribution, which can be defined as follows:

$$p(y|x, D) = N(y \mid \hat{\mu}, \sigma^{\wedge 2})$$

Here, the value of $\hat{\mu}$ and $\sigma^{\wedge 2}$ can be *analytically derived from the kernel, K* .

To summarize, *GP approximates the objective function by following a normal distribution assumption*. In practice, GP can also be utilized when we don't have zero mean processes, as per our previous assumption. However, we need to do some preprocessing on the values of the objective function to center them to zero. Choosing the *right covariance kernel, K*, is also crucial. It highly impacts the performance of our hyperparameter tuning process. The most popular kernel that's used in practice is the *Matern kernel*. However, we must choose the right kernel for our case, since each kernel has a characteristic that may or may not be suitable for our objective function. We will discuss the kernels that are available in the Scikit package in *Chapter 7, Hyperparameter Tuning via Scikit*.

The following table shows the list of pros and cons of BOGP compared to other variants of the BO hyperparameter tuning method:

Pros	Cons
Results in a posterior distribution that can be derived analytically, without the help of a numerical method.	Only works on continuous hyperparameters.
Relatively accurate approximation, even with a small number of observed points.	Not suitable for big hyperparameter spaces and a large number of samples.
	Can't exploit the parallel computing resources due to the sequential process of the method.

Figure 4.10 – Pros and cons of BOGP

In the previous section, we saw how GP works in practice, where we discussed the exploration and exploitation trade-off. You can revisit that example to get a better understanding of how GP works in practice through the help of visualizations.

In this section, we learned about utilizing GP as the surrogate model in BO, along with the pros and cons compared to other variants of BO. In the next section, we will learn about another variant of BO that utilizes random forest as the surrogate model.

Understanding SMAC

SMAC is part of the BO hyperparameter tuning method group and utilizes random forest as the surrogate model. This method is optimized to handle discrete or categorical hyperparameters. If your hyperparameter space is huge and is dominated by discrete hyperparameters, then SMAC is a good choice for you.

Similar to BOGP, SMAC also works by modeling the objective function. Specifically, it utilizes random forest as the surrogate model to create an estimation of the real objective function, which can then be passed to the acquisition function (see the *Introducing BO* section for more details).

Random forest is a **machine learning** (**ML**) algorithm that can be utilized in classification or regression tasks. It is built upon a collection of decision trees, which is known to perform well with categorical types of features. The name random forest comes from the fact that it is built from several decision trees. We will discuss random forest, along with its hyperparameters, in more detail in *Chapter 11, Understanding Hyperparameters of Popular Algorithms*.

The main difference between SMAC and BOGP lies in the type of surrogate model that's used in each method. While BOGP utilizes GP as the surrogate model, SMAC utilizes random forest as the surrogate model. The acquisition function that was used in the original paper on SMAC is the *EI function with some modifications* on how the optimization process in *Step 8* in the *Introducing BO* section is done, which also can be seen in the following screenshot:

7.   ...

8.   Sample the next set of hyperparameters by utilizing the acquisition function, *A*:

    I.   Perform optimization on the acquisition function, *A*, with the help of the surrogate model, *M*, to sample which hyperparameters are to be passed to the acquisition function.

    II.   Get the expected optimal set of hyperparameters based on the acquisition function, *A*.

9.   ...

Figure 4.11 – Optimization process of the acquisition function

In SMAC, similar to BOGP, we are also assuming that the distribution of our surrogate model's *prediction follows the Gaussian distribution*, as shown here:

$$p(y|x, D) = N(y \mid \hat{\mu}, \sigma^{\hat{}2})$$

Here, the $\hat{\mu}$ and $\sigma^{\hat{}2}$ values are derived from the *random forest prediction's mean and variance*, respectively.

We can also *utilize random forest to perform hyperparameter tuning on a random forest model*! How is this possible? How can a model be used to improve the performance of another model of the same type?

It is possible because we are treating one model as the surrogate model while the other one is the actual model that is fitted to the independent variables to predict the dependent variable. As the surrogate model, random forest will act as the regressor, which has the goal of learning the relationship between the hyperparameter space and the corresponding objective function. So, when we said that we are utilizing random forest to perform hyperparameter tuning on a random forest model, there are two random forest models with different goals and different input-output pairs!

Take a look at the following steps to get a better understanding of this concept. Note that the following procedure replaces *Steps 7* to *11* in the *Introducing BO* section:

6.   (The first few steps are the same as we saw earlier).

7.   Fit the *first random forest model*, which acts as a surrogate model, *M*, using the value pairs in *D*. Remember that *D* consists of pairs of hyperparameter values and the cross-validation score.

8. Sample the next set of hyperparameters by utilizing the acquisition function, *A*:

    I. Perform optimization on the acquisition function with the help of the surrogate model, *M*, to sample which hyperparameters are to be passed to the acquisition function.

    II. Get the optimal set of hyperparameters based on the acquisition function.

9. Compute the cross-validation score using the objective function, *f*, based on the output from *Step 8*. Note that the cross-validation score is computed based on the *second random forest model*, whose goal is to learn the relationship between the dependent and independent variables from our original problem.

10. Add the hyperparameters and cross-validation score pair from *Step 8* and *Step 9* to set *D*.

11. Repeat *Steps 7* to *10* until the stopping criteria are met.

12. (The last few steps are the same as we saw earlier).

You may be wondering, why bother utilizing the same ML algorithm as the surrogate model? Why don't we just perform a grid search or random search instead? Remember that the surrogate model is just one piece of the full BO algorithm. There is also the acquisition function and other optimization steps that can help us get the optimal set of hyperparameters faster. It is worth noting that *we can utilize any ML model* other than random forest. When it comes to tree-based ML models, XGBoost, CatBoost, and LightGBM are also popular among data scientists since they work well in practice.

In the *Introducing BO* section, we saw how GP works with the EI acquisition function to estimate a dummy objective function. Let's use the same dummy objective function, as defined here, and see the result of utilizing random forest (not necessarily the SMAC algorithm) as the surrogate model instead of GP. We will still use EI as the acquisition function in this example and the Scikit-optimize package as the implementation:

$$f(x) = \cos(6x) \cdot (1 - \sin(x5)) + \varepsilon$$

Here, ε is a noise that follows the standard normal distribution. Please see *Figure 4.2* for a visualization of this dummy objective function.

Let's set the number of trials and the exploitation versus exploration trade-off controller, δ, using the default values given by the Scikit-optimize package for the random forest surrogate model, which are 100 and 0.01, respectively. You can see how the random forest surrogate model fitting process works for the first five trials in the following figure:

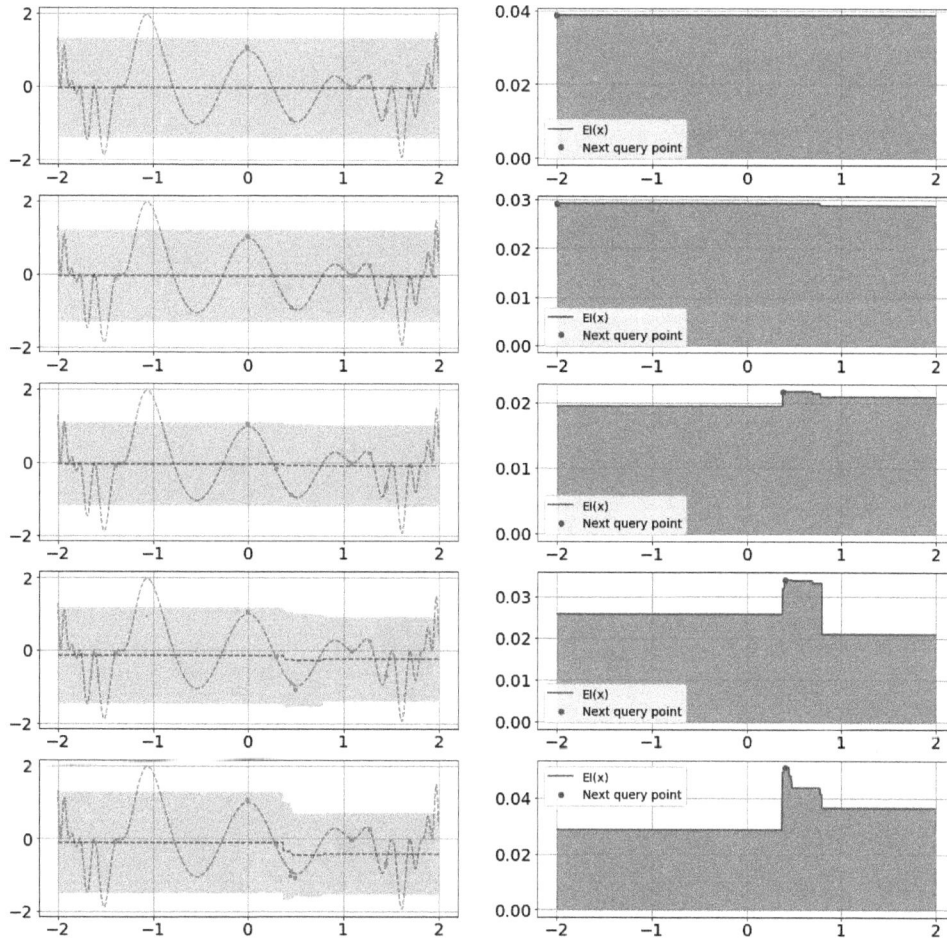

Figure 4.12 – Random forest and EI illustration; δ = 0.01; trials 1 – 5

As you can see, not many things happened in the first five trials. Even the approximation of the objective function that's given by the random forest (*see the green-dashed line*) is still very bad since it is just a straight line! Let's see what the condition is during trials 71 until 75:

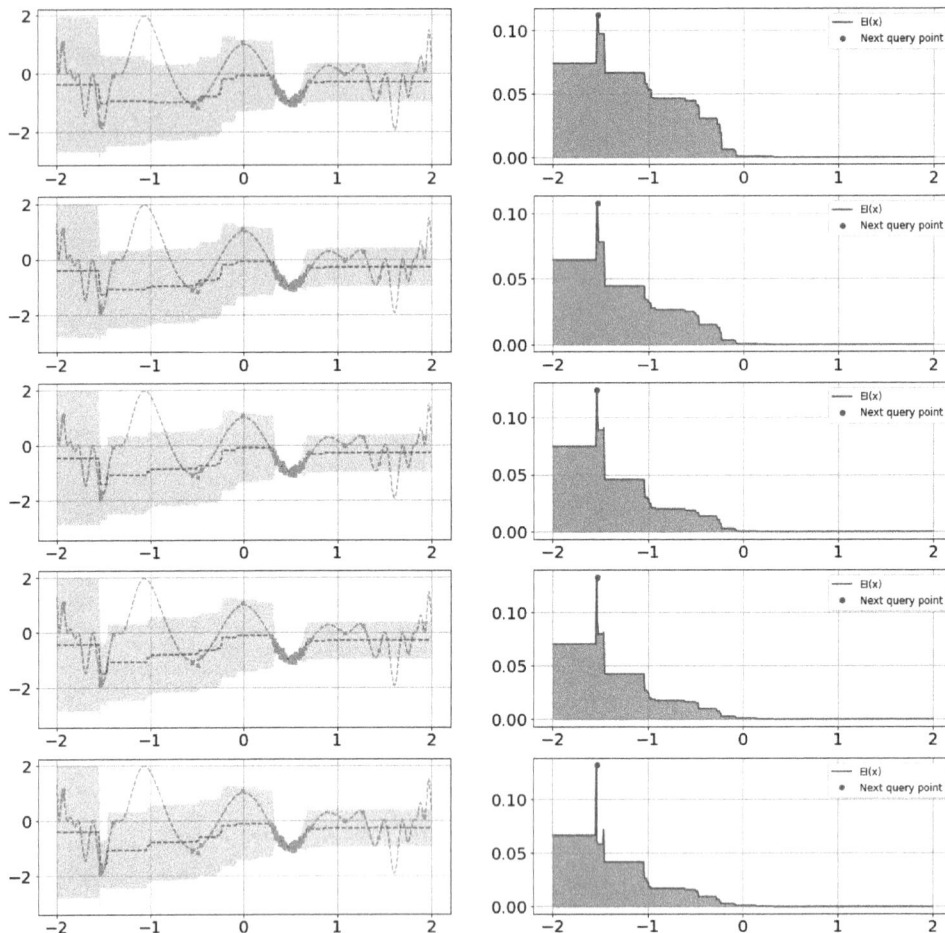

Figure 4.13 – Random forest and EI illustration; δ = 0.01; trials 71- 75

Here, we can see that our random forest surrogate model has improved a lot in estimating the true objective function. One interesting point is that the acquisition function curve looks very different from the one we saw when utilizing GP as the surrogate model. Here, the acquisition function looks edgier, just like the one we usually see from visualizing random forest. Finally, let's see what the final form of the approximated function is:

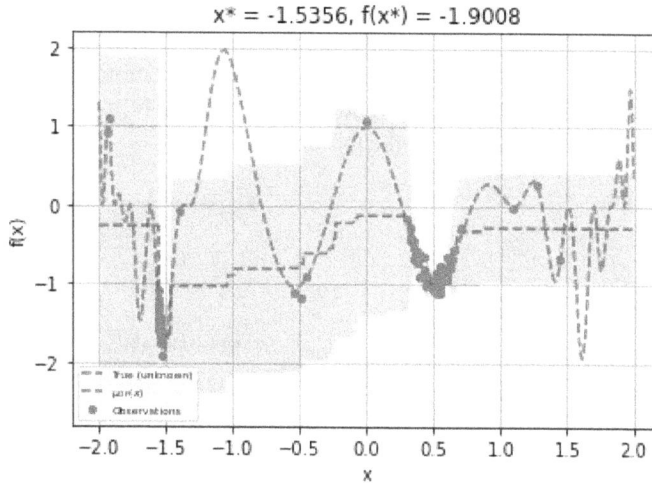

Figure 4.14 – Result after 100 trials; δ = 0.01

Here, we can see that random forest fails to fit the true objective function in general, but it succeeds to focus on the local minima of the objective function. This happens because *random forest needs a lot of data*, or in this case, the observed points (*see red dots*), to have a good approximation of the objective function. You can also see the convergence plot of the fitting process, starting from the first until the last trial, in the following plot. If we compare *Figure 4.15* to *Figure 4.5*, we can easily see that, in this example, random forest, when supported by the EI acquisition function, learns much slower than GP supported by EI:

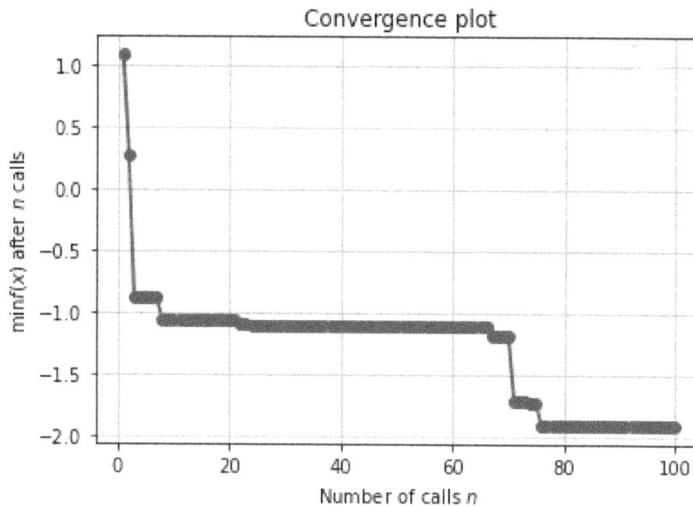

Figure 4.15 – Convergence plot

From *Figure 4.14*, we can also see that, currently, we are only focusing on several ranges and missing the global minima of the dummy objective function, which is located around the 1.5,1.75 range. Let's see if changing the value of δ to 100 can solve this issue. The expectation is that the EI acquisition function can help the random forest surrogate model *explore more* in other ranges of values as well. You can see the result of the first five trials in the following figure:

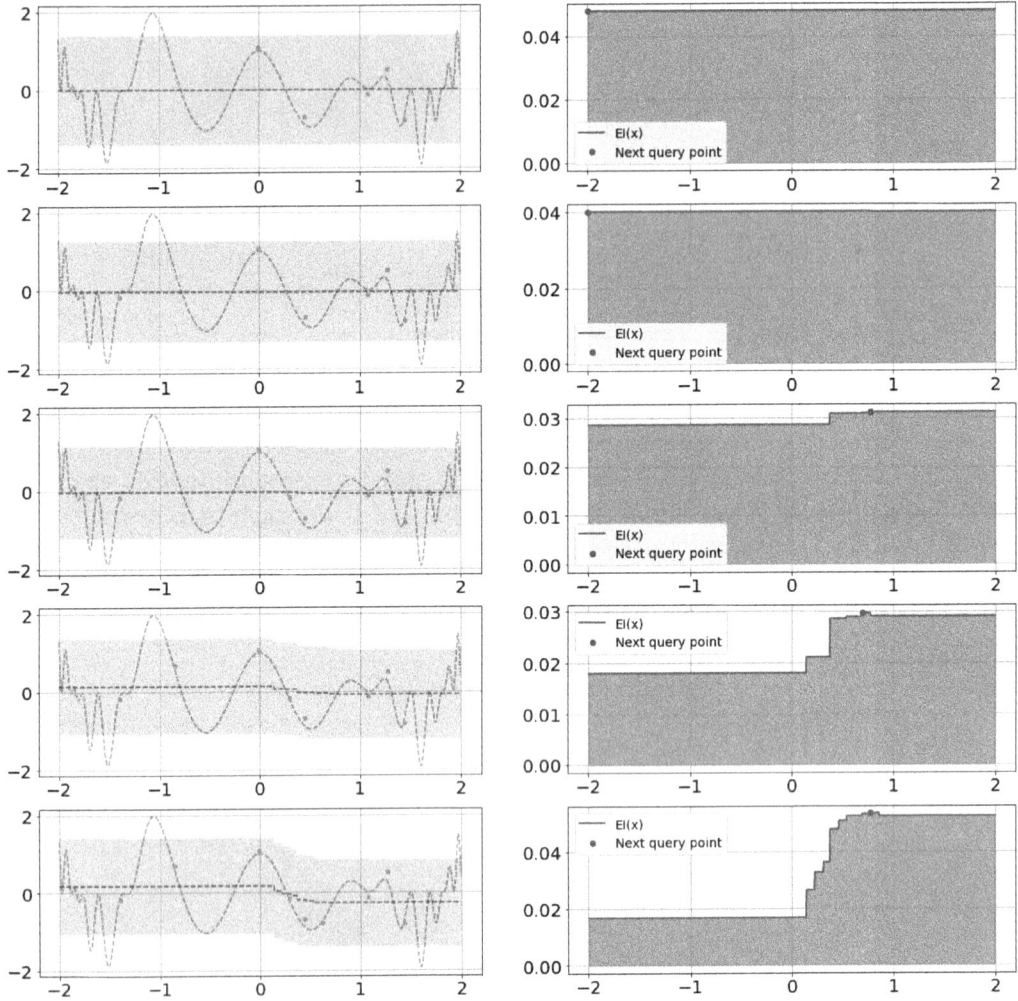

Figure 4.16 – Random forest and EI illustration; $\delta = 100$; trials 1 – 5

Similar to the first five trials of the default δ value, we still can't see much of the learning process. Let's see what the condition is during trials 71 until 75:

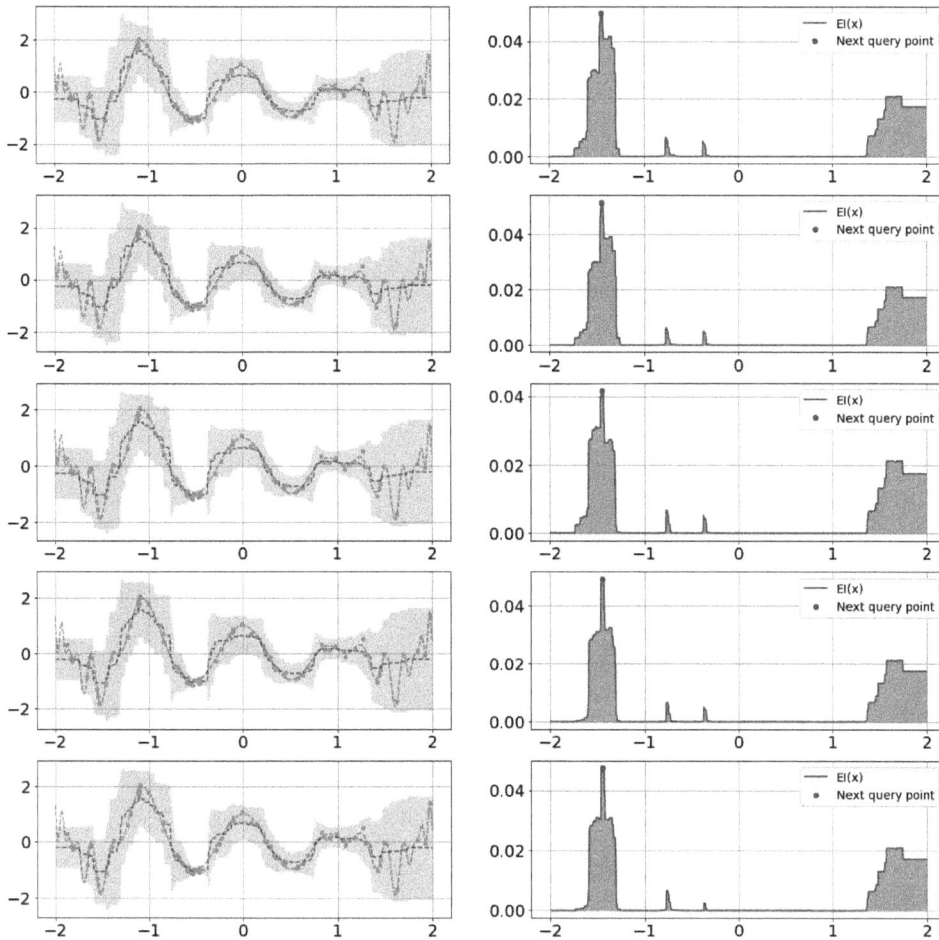

Figure 4.17 – Random forest and EI illustration; $\delta = 100$; trials 71 – 75

Here, we can see a very big difference between *Figure 4.17* and *Figure 4.13*. Finally, let's see what the final form of the approximated function is:

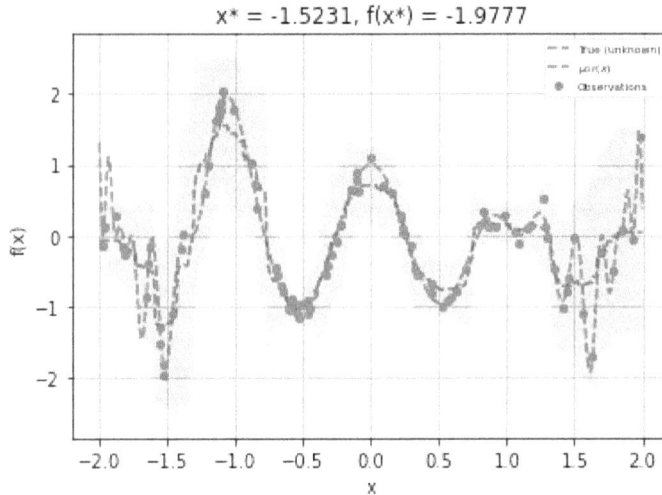

Figure 4.18 – Result after 100 trials; δ = 100

By changing the value of δ to 100, it seems that our expectation has been achieved. The approximation from the random forest surrogate model (*see the green-dashed line*) is now focusing on more than specific ranges. Moreover, we even get a better result compared to GP (see *Figure 4.4*). Again, it is worth noting that this is not always the case – you must experiment a lot on your own since different data, different objective functions, a different hyperparameter space, and different implementations may result in different conclusions. We will learn how to implement random forest as the surrogate model and how to produce these figures in *Chapter 7, Hyperparameter Tuning via Scikit*.

There is another method, called **Bayesian optimization inside a Grove (BOinG)**, whose goal is to get the best of both worlds by utilizing random forest and GP as surrogate models.

Bayesian Optimization Inside a Grove

See the following paper for more information: *Searching in the Forest for Local Bayesian Optimization*, by Difan Deng and Marius Lindauer (`https://arxiv.org/abs/2111.05834`).

BOinG works by using *two-stage optimization* by using global and local models to cut down the computational cost and focus more on the promising subspace, respectively. In BOinG, random forest is utilized as the global model and GP as the local model. The global model is responsible for searching the promising subspace of the local model. Thus, a global model should be flexible enough to handle complex problems with different types of hyperparameters. Since the local model only searches in a promising subspace, it is possible to utilize a more accurate but expensive model, such as GP.

The following table lists the pros and cons of utilizing random forest as a surrogate model compared to other variants of the BO hyperparameter tuning method:

Pros	Cons
Handles categorical features very well.	Poor estimator on the space with a lower number of observed points.
Suitable for big hyperparameter spaces and a large number of samples.	Can't exploit the parallel computing resources due to the sequential process of the method.
Able to handle conditional hyperparameters.	

Figure 4.19 – Pros and cons of utilizing random forest as a surrogate model

A conditional hyperparameter is a hyperparameter that will only be utilized when a certain condition is met. The tree structure of random forest is very suitable for this kind of situation since it can just add another branch of the tree to check whether the condition is met or not. The condition is usually just a specific value or range of other hyperparameters in the space.

Now that you are aware of SMAC and utilizing random forest as a surrogate model in general, in the next section, we will discuss another variant of BO that has a different approach in terms of approximating the objective function.

Understanding TPE

TPE is another variant of BO that performs well in general and can be utilized for both categorical and continuous types of hyperparameters. Unlike BOGP, which has cubical time complexity, TPE runs in linear time. TPE is suggested if you have a huge hyperparameter space and have a very tight budget for evaluating the cross-validation score.

The main difference between TPE and BOGP or SMAC is in the way that it models the relationship between hyperparameters and the cross-validation score. Unlike BOGP or SMAC, which approximate the value of the objective function, or the posterior probability, $p(y|x)$, *TPE works the other way around*. It tries to get the optimal hyperparameters based on the condition of the objective function, or the likelihood probability, $p(x|y)$ (see the explanation of Bayes Theorem in the *Understanding BO GP* section).

In other words, unlike BOGP or SMAC, which construct a predictive distribution over the objective function, TPE tries to utilize the information of the objective function to *model the hyperparameter distributions*. To be more precise, when the optimization problem is in the form of a *minimization problem*, $p(x|y)$ is defined as follows:

$$p(x|y) = l(x) \; if \; y < y^* \; and \; g(x) \; if \; y \geq y^*$$

Here, $l(x)$ and $g(x)$ are utilized when the value of the objective function is lower or higher than the threshold, y^*, respectively. There is no specific rule on how to choose the threshold, y^*. However, in the **Hyperopt** and **Microsoft NNI** implementations, this threshold is chosen based on the TPE's hyperparameter, γ, and the number of observed points in D up to the current trial. The definition of $p(x|y)$ tells us that TPE has two models that act as the learning algorithm based on the value of the objective function, which is ruled by the threshold, y^*.

When the *distribution of hyperparameters is continuous*, TPE will utilize **Gaussian mixture models** (**GMMs**), along with the EI acquisition function, to suggest the next set of hyperparameters to be tested. If the continuous distribution is not a Gaussian distribution, then TPE will convert it to mimic the Gaussian distribution. For example, if the specified hyperparameter distribution is the uniform distribution, then it will be converted into a truncated Gaussian distribution.

The probabilities of the different possible outcomes for the multinomial distribution within the GMM, and the mean and variance values for the normal distribution within the GMM, are generated by the **adaptive Parzen estimator**. This estimator is responsible for constructing the two probability distributions, $l(x)$ and $g(x)$, based on the mean and variance of the normal hyperparameter distribution, as well as the hyperparameter value of all observed points in D up to the current trial.

When the *distribution is categorical or discrete*, TPE will convert the categorical distribution into a re-weighted categorical and use *weighted random sampling*, along with the EI acquisition function, to suggest the expected best set of hyperparameters. The weights in the random sampling procedure are generated based on the historical counts of the hyperparameter value.

The EI acquisition function definition in TPE is a bit different from the definition we learned about in the *Introducing BO* section. In TPE, we are using Bayes Theorem when deriving the EI formula. The simple formulation of the EI acquisition function in TPE is defined as follows:

$$EI(x) \propto \frac{l(x)}{g(x)}$$

The proportionality defined here tells us that to get a high value of EI, we need to get a high $\frac{l(x)}{g(x)}$ ratio. In other words, when the optimization problem is in the form of a *minimization problem*, the EI acquisition function must suggest more hyperparameters from $l(x)$ over $g(x)$. It is the other way around when the optimization problem is in the form of a *maximization problem*. For example, when we use accuracy to measure the performance of our classification model, then we should sample more hyperparameters from $g(x)$ over $l(x)$.

To summarize, TPE works as follows. Note that the following procedure describes how TPE works for the *minimization problem*. This procedure replaces *Steps 7 to 11* in the *Introducing BO* section:

6. (The first few steps are the same as we saw earlier).

7. Divide pairs of hyperparameter values and cross-validation scores in *D* into two groups based on the threshold, y^*, namely *below* and *above* groups (see *Figure 4.19*).

8. Sample the next set of hyperparameters by utilizing the EI acquisition function:

   I. For each group, calculate the probabilities, means, and variances for the GMM using the adaptive Parzen estimator (if it's a continuous type) or weights for random sampling (if it's a categorical type).

   II. For each group, fit the GMM (if it's a continuous type), or perform random sampling (if it's a categorical type), to sample which hyperparameters will be passed to the EI acquisition function.

   III. For each group, calculate the probability of those samples being good samples (for the below group), or the probability of those samples being bad samples (for the above group).

   IV. Get the expected optimal set of hyperparameters based on the EI acquisition function.

9. Compute the cross-validation score using the objective function, *f*, based on the output from *Step 8*.

10. Add the hyperparameters and cross-validation score pair from *Step 8* and *Step 9* to set *D*.

11. Repeat *Steps 7 to 10* until the stopping criteria have been met.

12. (The last few steps are the same as we saw earlier):

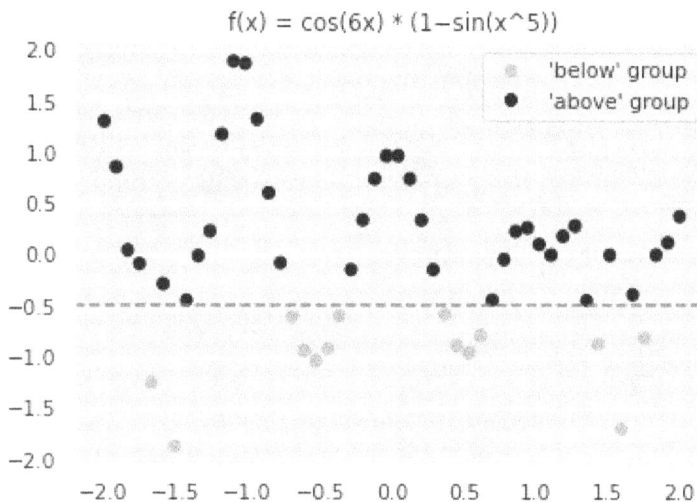

Figure 4.20 – Illustration of groups division in TPE

Based on the stated procedure and the preceding plot, we can see that, unlike BOGP or SMAC, which constructs a predictive distribution over the objective function, TPE tries to utilize the information of the objective function to model the hyperparameter distributions. This way, we are not only focusing on the best-observed points during the trials – we are focusing on the *distribution of the best-observed points* instead.

You may be wondering why the *Tree-structured* term is within the TPE method's name. This term refers to the conditional hyperparameters that we discussed in the previous section. This means that there are hyperparameters in the space that will only be utilized when a certain condition is met. We will see what a tree-structured or conditional hyperparameter space looks like in *Chapter 8, Hyperparameter Tuning via Hyperopt*, and *Chapter 9, Hyperparameter Tuning via Optuna*.

One of the drawbacks that TPE has is that it may *overlook the interdependencies among hyperparameters* in a certain space since the Parzen estimators work univariately. However, this is not the case for BOGP or SMAC, since the surrogate model is constructed based on the configurations in the hyperparameter space. Thus, they can take into account the interdependencies among hyperparameters. Fortunately, there is an implementation of TPE that overcomes this drawback. The **Optuna** package provides the **multivariate TPE** implementation, which can take into account the interdependencies among hyperparameters.

The following table lists of pros and cons of utilizing TPE compared to other variants of the BO hyperparameter tuning method:

Pros	Cons
Performs well in general.	Can't capture interdependencies between hyperparameters, though this can be solved by multivariate TPE.
Suitable for big hyperparameter spaces and a large number of samples.	Can't exploit the parallel computing resources because of the sequential process of the method.
Can handle categorical and continuous types of hyperparameters.	
Can handle conditional hyperparameters.	

Figure 4.21 – Pros and cons of TPE

> **Important Note**
>
> Some implementations support parallel tuning, but with a trade-off between the suggested hyperparameter quality and the wall time. The Microsoft NNI package supports this feature via the `constant_liar_type` argument, which will be discussed in more detail in *Chapter 10, Advanced Hyperparameter Tuning with DEAP and Microsoft NNI*.

In this section, we learned about TPE, along with its pros and cons compared to other variants of BO. In the next section, we will learn about another variant of BO that has a slightly modified algorithm compared to the BO method in general.

Understanding Metis

Metis is one of the variants of BO that has several algorithm modifications compared to the BO method in general. Metis utilizes GP and GMM in its algorithm. GP is used as the surrogate model and outliers detector, while GMM is used as part of the acquisition function, similar to TPE.

What makes Metis different from other BO methods, in general, is that it can *balance exploration and exploitation more data-efficiently* than the EI acquisition function. It can also *handle noise in the data that doesn't follow the Gaussian* distribution, and this is the case most of the time. Unlike most of the methods that perform random sampling to initialize the set of hyperparameters and cross-validation score, *D*, Metis utilizes **Latin Hypercube Sampling** (**LHS**), which is a stratified sampling procedure based on the equal interval of each hyperparameter. This sampling method is believed to be more data-efficient compared to random sampling to achieve the same exploration coverage.

So, how can Metis balance exploration and exploitation more efficiently than the EI acquisition function, in terms of the needs of the observed points? This is achieved through the *custom acquisition function* that Metis has, which consists of three sub-acquisition functions, as shown here:

- **Lowest confidence** (**LC**): This sub-acquisition function's goal is to sample hyperparameters with the highest uncertainty. In other words, the goal of this sub-acquisition function is to *maximize exploration*. This function is defined as follows:

$$LC(x) = -\frac{1.96 \cdot 2 \cdot \sigma(x)}{\mu(x)}$$

- **Parzen estimator**: This sub-acquisition function is inspired by the TPE method, which utilizes GMM to estimate how likely the sampled hyperparameter is part of the *below* or *above* group (see the *Understanding TPE* section for more details). The goal of this sub-acquisition function is to sample hyperparameters with the highest probability to be the optimum hyperparameters. In other words, it is *optimized for exploitation*.

- **Outliers detector**: As its name suggests, the goal of this sub-acquisition function is to *detect outliers within D*. The detected outlier will then be suggested as the candidate to be resampled in the next trial. Metis utilized GP to build the outliers detector or the **diagnostic model**. This diagnostic model works by comparing each of the cross-validation scores in *D* with the mean and standard deviation estimated by the GP. If the absolute difference between the cross-validation score and the estimated mean is greater than some constant multiplied by the estimated standard deviation, then it is flagged as an outlier. In other words, the diagnostic model will mark the hyperparameter as an outlier if it *lies outside the confidence interval of the GP estimation*. The constant for the 98% confidence interval is 2.326.

Based on candidates suggested by these three sub-acquisition functions, Metis will then compute their *information gain* to select the final candidate to be included in the next trial. This selection process is done by utilizing the lower bound of the GP estimation confidence interval. Metis will measure the difference between the lower bound of the interval and the expected mean from GP. The candidate that has the highest improvement will be selected as the final candidate.

It is worth noting that Metis can handle non-Gaussian noise in the data because of the diagnostic model. The detected outliers made it possible for Metis to resample the previously tested hyperparameters so that it is robust to non-Gaussian noise as well. This way, Metis can *balance exploration, exploitation, and re-sampling* during the hyperparameter tuning process.

To have a better understanding of how Metis works, take a look at the following procedure. Note that the following procedure replaces *Steps 6 to 11* in the *Introducing BO* section.

5. (The first few steps are the same as we saw earlier).

6. Initialize several pairs of hyperparameter values and cross-validations scores using the LHS method, and store them in *D*.

7. Fit a GP that acts as a surrogate model, *M*, using the value pairs in *D*.

8. Sample the next set of hyperparameters by utilizing the *custom acquisition function*, which consists of three sub-acquisition functions:

    I.   Get the current best optimum set of hyperparameters

    II.   Get the suggested hyperparameters for *exploration* via the LC sub-acquisition function

    III. Get the suggested hyperparameters for *exploitation* via the Parzen estimator

    IV.   Get the suggested hyperparameters to be resampled based on the *detected outliers* by the diagnostic model.

    V.   Calculate the *information gain* from each suggested candidate.

    VI.   Select the candidate that has the highest information gain.

    VII. If no candidate is suggested, then pick one random candidate.

9. Compute the cross-validation score using the objective function, *f*, based on the output from *Step 8*. Note that the cross-validation score is computed based on the *second random forest model*, whose goal is to learn the relationship between the dependent and independent variables from our original problem.

10. Add the hyperparameters and cross-validation score pair from *Step 8* and *Step 9* to set *D*.

11. Repeat *Steps 7* to *10* until the stopping criteria are met.

12. (*The last few steps are the same as we saw earlier*).

The following table lists the pros and cons of utilizing Metis compared to other variants of the BO hyperparameter tuning method:

Pros	Cons
Balances exploration and exploitation more data-efficiently.	Only works on numerical hyperparameters.
Can handle non-Gaussian noise in the data.	Not suitable for big hyperparameter spaces and a lot of experimental trials.
	Can't exploit the parallel computing resources because of the sequential process of the method.

Figure 4.22 – Pros and cons of Metis

It is also worth noting that, unlike other BO variants, there is only one package that implements Metis for the hyperparameter tuning method, which is **Microsoft NNI**. As you may have noticed, all the variants of BO that were discussed in this chapter have the drawback of not being able to exploit parallel computing resources. So, why didn't we put that drawback in the first section instead? Because there is a variant of BO, namely BOHB, that can exploit the parallel computing resources. We will discuss BOHB in more detail in *Chapter 6, Exploring Multi-Fidelity Optimization*.

In this section, we covered Metis in detail, including, what it is, how it works, what makes it different from other BO variants, and its pros and cons.

Summary

In this chapter, we discussed the second out of four groups of hyperparameter tuning methods, called the BO group. We not only discussed BO in general but also several of its variants, including BOGP, SMAC, TPE, and Metis. We saw what makes each of the variants differ from each other, along with the pros and cons of each. At this point, you should be able to explain BO with confidence when someone asks you and apply hyperparameter tuning methods in this group with ease.

In the next chapter, we will start discussing heuristic search, the third group of hyperparameter tuning methods. The goal of the next chapter is similar to this chapter: to provide a better understanding of the methods that belong to the heuristic search group.

5
Exploring Heuristic Search

Heuristic search is the third out of four groups of hyperparameter tuning methods. The key difference between this group and the other groups is that all the methods that belong to this group work by performing *trial and error* to achieve the optimal solution. Similar to the acquisition function in Bayesian optimization (see *Chapter 4, Exploring Bayesian Optimization*), all methods in this group also employ the concept of *exploration versus exploitation*. **Exploration** means performing a search in the unexplored space to lower the probability of being stuck in the local optima, while **exploitation** means performing a search in the local space that is known to have a good chance of containing the optimal solution.

In this chapter, we will discuss several methods that belong to the heuristic search group, including **simulated annealing (SA)**, **genetic algorithms (GAs)**, **particle swarm optimization (PSO)**, and **Population-Based Training (PBT)**. Similar to *Chapter 4*, we will discuss the definition of each method, what the differences are between them, how they work, and the pros and cons of each method.

By the end of this chapter, you will understand the concept of the aforementioned hyperparameter tuning methods that belong to the heuristic search group. You will be able to explain these methods with confidence when someone asks you, at both a high-level and detailed fashion, along with the pros and cons. Once you are confident enough to explain them to other people, this means you have understood the ins and outs of each method. Thus, in practice, you can understand what's happening if there are errors or you don't get the expected results; you will also know how to configure the method so that it matches your specific problem.

In this chapter, we will cover the following topics:

- Understanding simulated annealing
- Understanding genetic algorithms
- Understanding particle swarm optimization
- Understanding population based training

Understanding simulated annealing

SA is the heuristic search method that is inspired by the process of **metal annealing** in metallurgy. This method is similar to the random search hyperparameter tuning method (see *Chapter 3, Exploring Exhaustive Search*), except for the existence of a criterion that guides how the hyperparameter tuning process works. In other words, SA is like a *smoothed version of random search*. Just like random search, it is suggested to use SA when each trial doesn't take too much time and you have enough computational resources.

In the metal annealing process, the metal is heated to a very high temperature for a certain time and slowly cooled to increase its strength, reducing its hardness and making it easier to work with. The goal of giving a very high heat is to excite the metal's atoms so that they can move around freely and randomly. During this random movement, atoms usually tend to form a better configuration. Then, the slow cooling process is performed so that we can have a crystalline form of the material.

Just like in the metal annealing process, SA works by randomly choosing the set of hyperparameters to be tested. At each trial, the method will consider some of the "neighbors" of the current set, randomly. If the acceptance criterion is met, then the method will change its focus to that "neighbor" set. The acceptance criterion is not a deterministic function, it is a stochastic function, which means probability comes into play during the process. This probabilistic way of deciding is similar to the cooling phase in the metal annealing process, where we accept a smaller number of bad hyperparameter sets as more parts of the search space are explored.

SA is a modified version of one of the most popular heuristic optimization methods, known as **stochastic hill climbing (SHC)**. SHC is very simple to understand and implement, which means that SA is as well. *In general*, SHC works by initializing the random point within a pre-defined bound (the *hyperparameter space, in our case*) and treating it as the current best solution. Then, it randomly searches for the next candidate within the surrounding of the selected point. Then, we need to compare the selected candidate with the current best solution. If the candidate is better than or equal to the current best solution, SHC will treat the candidate as the new best solution. This process is repeated until the stopping criterion is met.

The following steps show how SHC optimization works in general:

1. Define the bound of the space, *B*, and the step size, *S*.

2. Define the stopping criterion. Usually, it is defined as the number of iterations, but other stopping criteria definitions also work.

3. Initialize the random point within the bound, *B*.

4. Set the selected point from *Step 3* as the current point, *current_point*, as well as the best point, *best_point*.

5. Randomly sample the next candidate within the *S* distance from *best_point* and within the bound, *B*, then store it as *candidate_point*.

6. If *candidate_point* is better than or equal to *best_point*, then replace *best_point* with *candidate_point*.

7. Replace *current_point* with *candidate_point*.

8. Repeat *Steps 5* to *7* until the stopping criterion is met.

The main difference between SA and SHC is located in *Steps 5* and *6*. In SHC, we always sample the next candidate from the surrounding of the *best_point*, while in SA, we sample from the surrounding of *current_point*. In SHC, we only accept a candidate that is better than or equal to the current best solution, while in SA, we *may also accept a worse candidate* with a certain probability that is guided by the acceptance criterion, *AC*, which is defined as follows:

$$AC(T, \Delta f) = \exp\left(-\frac{\Delta f}{T}\right) \; if \; candidate\_point \; worse \; than \; current\_point \; else \; 1$$

Here, $\Delta f = |f(candidate\_point) - f(current\_point)|$, f is the objective function and T is *temperature with a positive value*. See *Chapter 4* if you are not familiar with the objective function term.

The *AC* formula results in a value between 0 and 1, where it always results in a value of 1 when the *candidate_point* is better than or equal to the *current_point*. In other words, we always accept the *candidate_point* when it is better than or equal to the *current_point*. It is worth noting that *better* does not necessarily mean has a greater value. If you are working with a maximization problem, then better means greater. However, if you are working with a minimization problem, then it is the other way around. For example, if the cross-validation score you are measuring is the **mean squared error** (**MSE**), where a lower score corresponds to better performance, then the *candidate_point* is considered better than the *current_point* if the $f(candidate\_point)$ value is less than $f(current\_point)$.

Although *AC* is impacted by both Δf and T, we can only control the value of T. In practice, the *initial value* of T *is treated as a hyperparameter* and is usually set to a high value. Over the number of trials, the value of T is decreased following the so-called **annealing schedule** or cooling schedule scheme. There are several annealing schedule schemes that we can follow. The three most popular schemes are as follows:

- **Geometric cooling**: This annealing schedule works by decreasing the temperature via a cooling factor of $0 < \alpha < 1$. In geometric cooling, the initial temperature, T_0, is multiplied by the cooling factor *iter* number of times, where *iter* is the current number of iterations:

$$T = \alpha^{iter} \cdot T_0$$

This can be seen in the following graph:

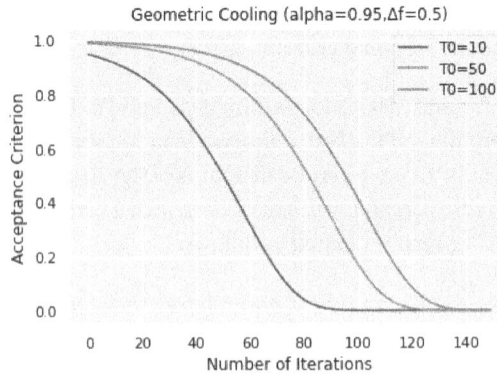

Geometric Cooling (alpha=0.95,Δf=0.5)

Figure 5.1 – Effect of the initial temperature in geometric cooling on the acceptable criterion

- **Linear cooling**: This annealing schedule works by decreasing the temperature linearly via a cooling factor, β. The value of β is chosen in such a way that T will still have a positive value after *iter* iterations. For example, $\beta = \frac{(T_0 - T_f)}{t_f}$, where T_f is the expected final temperature after t_f iterations:

$$T = T_0 - \beta \cdot iter$$

The following graph shows this annealing schedule:

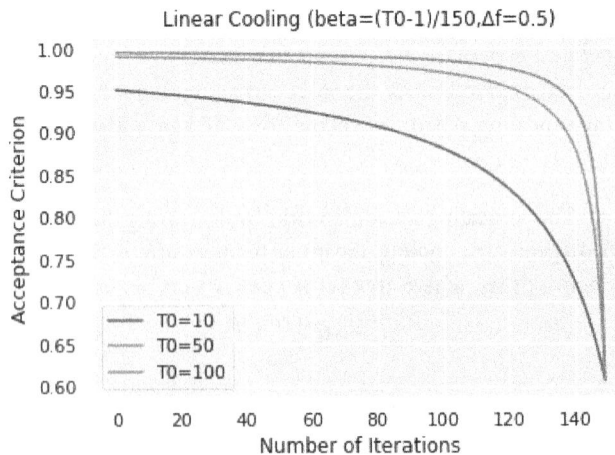

Linear Cooling (beta=(T0-1)/150,Δf=0.5)

Figure 5.2 – Effect of the initial temperature in linear cooling on the acceptable criterion

- **Fast SA**: This annealing schedule works by decreasing the temperature proportional to the current number of iterations, $iter$:

$$T = \frac{T_0}{iter}$$

This annealing schedule can be seen in the following graph:

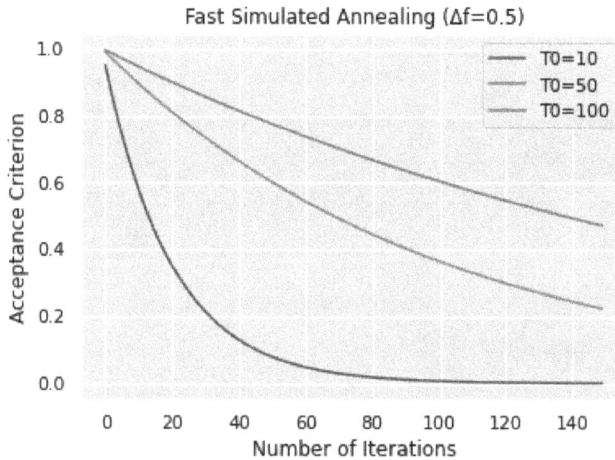

Figure 5.3 – Effect of the initial temperature in fast SA on the acceptable criterion

Based on *Figures 5.1* to *5.3*, we can see that no matter what annealing schedule scheme we use and what the initial temperature is, we will always have a lower AC value as the number of iterations increases, which means we will *accept fewer bad candidates as the number of iterations increases*. However, why do we need to accept bad candidates in the first place? The main purpose of SA not directly rejecting worse candidates, as in the SHC method, is to *balance the exploration and exploitation trade-off*. The high initial value of temperature allows SA to explore most of the parts of the hyperparameter space, and slowly focus on specific parts of the space as the number of iterations increases, just like how the metal annealing process works.

Remember that AC only takes Δf into account when the *candidate_point* is worse than the *current_point*. This means that, based on *Figure 5.4*, we can say that the worse the suggested candidate is (the higher Δf is), the lower the value of AC will be, and thus, the lower the probability of accepting the suggested bad candidate. This is the other way around for T in that the higher the value of T is, the higher the value of AC will be, and thus, the higher the probability of accepting the suggested bad candidate (see *Figures 5.1* to *5.3*):

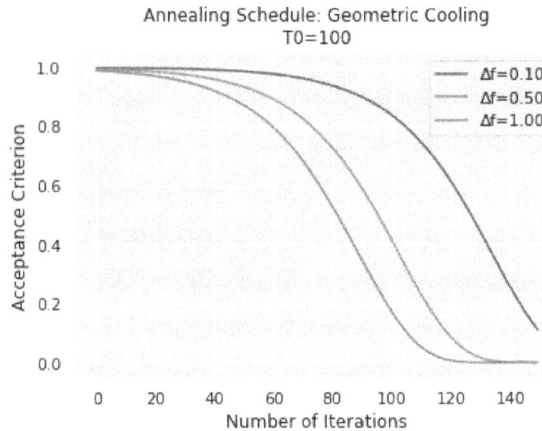

Figure 5.4 – Effect of Δf on the acceptable criterion

To summarize, the following steps show how *SA works as a hyperparameter tuning method*:

1. Split the original full data into train and test sets (see *Chapter 1, Evaluating Machine Learning Models*).

2. Define the hyperparameter space, H, with the accompanied distributions.

3. Define the initial temperature, $T0$.

4. Define the objective function, f, based on the train set (see *Chapter 4*).

5. Define the stopping criterion. Usually, the number of trials is used. However, it is also possible to use the time taken or convergence as the stopping criterion.

6. Set the current temperature, T, using the value from $T0$.

7. Initialize a random set of hyperparameters that have been sampled from the hyperparameter space, H.

8. Set the selected set from *Step 7* as the current set, *current_set*, as well as the best set, *best_set*.

9. Randomly sample the next candidate set, *candidate_set*, from the "neighbor" of the *current_set* within the hyperparameter space, H. The definition of the "neighbor" may differ across different types of hyperparameter distributions.

10. Generate a random number between 0 and 1 from the uniform distribution and store it as *rnd*.

11. Decide whether to accept the *candidate_set* or not:

    I.   Calculate the value of AC using the value of T, *f(candidate_set)*, and *f(current_set)*.

    II.   If the value of *rnd* is smaller than AC, then replace *current_set* with *candidate_set*.

    III.   If *candidate_set* is better than or equal to *current_set*, then replace *best_set* with *candidate_set*.

12. Apply the annealing schedule to the temperature, T.

13. Repeat *Steps 9* to *12* until the stopping criterion is met.

14. Train on the full training set using the *best_set* hyperparameters.

15. Evaluate the final trained model on the test set.

The following table lists the list of pros and cons of SA as a hyperparameter tuning method:

Pros	Cons
It has all the pros that random search has (see *Chapter 3*).	It has all the cons that random search has (see *Chapter 3*).
Ability to focus more on parts of the search space that are more likely to contain optimal hyperparameters.	May skip parts of the search space that contain optimal hyperparameters.
	Higher computational cost due to the need of calculating AC in each trial.

Figure 5.5 – Pros and cons of SA

In this section, we learned about SA, starting from what it is, how it works, what makes it different from SHC and random search, and its pros and cons. We will discuss another interesting heuristic search method that is inspired by the natural selection theory in the next section.

Understanding genetic algorithms

GAs are popular heuristic search methods that are inspired by Charles Darwin's *theory of natural selection*. Unlike SA, which is classified as a **single-point-based** heuristic search method, GAs are categorized as **population-based** methods since they maintain a group of possible candidate solutions instead of just a single candidate solution at each trial. As a hyperparameter tuning method, you are recommended to utilize a GA when each trial doesn't take too much time and you have enough computational resources, such as parallel computing resources.

To have a better understanding of GAs, let's start with a simple example. Let's say we have a task to generate a pre-defined target word based on *only* a collection of words that are built from 26 alphabet letters in lowercase. For instance, the target word is "big," and we have a collection that consists of the words "sea," "pig," "dog," "bus," and "tie."

Based on the given collection of words, what should we do to generate the word "big?" It is no doubt a very easy and straightforward task. We just have to pick the letter "b" from the word "bus," "i" from the word "pig" or "tie," and "g" from the word "dog." Voila! We get the word "big." You may be wondering how this example is related to the GA method or even the natural selection theory. This example is a very simple task and there is no need to utilize a GA to solve the problem. However, we need this kind of example so that you have a better understanding of how GAs work since you already know the correct answer in the first place.

To solve this task using GA, you must know the three key items in GA related to the evolution theory. The first key item is **variation**. Imagine if the given collection of words consists of only the word "sea." There's no way we can generate the word "big" based on only the word "sea." This is why variation is needed in the **initial population** (*the collection of words, in our example*). Without enough variation, we may not be able to achieve the optimal solution (*to generate the word "big," in our example*) since there is no **individual** (*each word in the collection of words, in our example*) within the population that can evolve to the target word.

> **Important Note**
>
> Population is *not the hyperparameter space*. In GAs or other population-based heuristic search methods, population refers to the candidates of the optimal hyperparameter set.

The second key item is **selection**. You can think of this item as being similar to the idea of natural selection that happens in the real world. It's about selecting individuals that are more suitable for the surrounding environment (*words that are similar to the word "big," in our example*) and thus can survive in the world. In GAs, we need quantitative guidance for us to perform the selection, which is usually called the **fitness function**. This function helps us judge how good an individual is concerning the objective we want to achieve. In our example, we can create a fitness function that measures the proportion of indexes of the word that has the same letters as the target word in the corresponding indexes. For example, the word "tie" has a fitness score of $\frac{1}{3}$ since only one out of three indexes contains the same letters as the target word, which is the index one that has the letter "i."

Using this fitness function, we can evaluate the fitness score for each individual in the population, and then select which individuals should be added to the **mating pool** as **parents**. The mating pool is a collection of individuals that are considered high-quality individuals and thus called parents.

The third key item is **heredity**. This item refers to the concept of **reproduction** or passing parents' **genes** (*each letter in the word, in our example*) to their children or **offspring**. How is reproduction done in GAs? Taking the same spirit of natural selection, in Gas, we only perform the reproduction step from parents in the mating pool, meaning we only want to mate high-quality individuals with the hope to get only high-quality offspring in the next **generation** (a *new population is created in the next iteration*). There are two steps in the reproduction phase, namely the **crossover** and **mutation** steps. The crossover step is when we randomly mix or permute parents' genes to generate offspring's genes, while the mutation step is when we randomly change the value of offspring's genes to add variation to the genes (see *Figure 5.6*). An individual that is mutated is called a **mutant**. The random value that is used in the mutation step should be drawn from the same gene's distribution, meaning we can only use lower-case letters as the random values in our example, not floating points or integers:

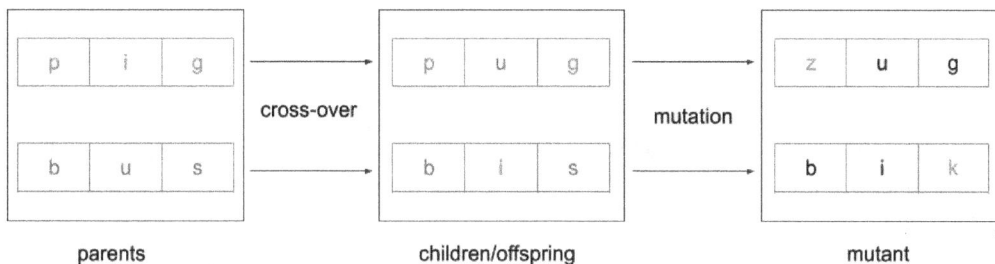

Figure 5.6 – The crossover and mutation steps in a GA

Now that you are aware of the three key items in a GA, we can start solving the task from the previous example using a GA. Let's assume we haven't been given the collection of words so that we can learn the complete procedures of the GA. The target word is still "big."

First, we must initialize a population with the *NPOP* number of individuals. The initialization process is usually done randomly to ensure we have enough variation in the population. By random, this means that the genes of each individual in the population are generated randomly. Let's say we want to generate the initial population, which consists of seven individuals, where the generated results are "bee," "tea," "pie," "bit," "dog," "cat," and "dig."

Now, we can evaluate the fitness score of each individual in the population. Let's say we use the fitness function that was defined previously. So, we got the following scores for each individual; "bee:" $\frac{1}{3}$, "tea:" 0, "pie:" $\frac{1}{3}$, "bit:" $\frac{2}{3}$, "dog:" $\frac{1}{3}$, "cat:" 0, and "dig:" $\frac{2}{3}$.

Based on the fitness score of each individual, we can select which individual should be added to the mating pool as a parent. There are many strategies that we can adopt to select the best individuals from the population, but in this case, let's just get the top three individuals based on the fitness score and randomly select individuals that have the same fitness score. Let's say that, after running the selection strategy, we get a mating pool that consists of "bit," "dig," and "bee" as parents.

The next step is to perform the crossover and mutation steps. Before that, however, we need to specify the crossover probability, *CXPB*, and the mutation probability, *MUTPB*, which defines the probability of crossing two parents in the mating pool and mutating an offspring, respectively. This means we are neither performing crossover on all parent pairs nor mutating all offspring – we will only perform those steps based on the predefined probability. Let's say that only "dig" and "bee" have chosen to be crossed, and the resulting offspring of the crossover is "deg" and "bie." So, the mating pool currently consists of "bit," "deg," and "bie." Now, we need to perform mutation on "deg" and "bie." Let's say that after mutating them, we got "den" and "tie." This means that the mating pool is currently consisting of "bit," "den," and "tie."

After performing the crossover and mutation steps, we need to generate a new population for the next generation. The new population will consist of all crossed parents, mutated offspring, as well as other individuals from the current population. So, the next population consists of "bit," "den," "tie," "tea," "pie," "dog," and "cat."

Based on the new population, we have to repeat the selection, crossover, and mutation process. This procedure needs to be done *NGEN* times, where NGEN refers to the number of generations, and it is predefined by the developer.

The following steps define *how GA works in general*, as an optimization method:

1. Define the population size, *NPOP*, the crossover probability, *CXPB*, the mutation probability, *MUTPB*, and the number of generations or number of trials, *NGEN*.

2. Define the fitness function, *f*.

3. Initialize a population with *NPOP* individuals, where each individual's genes are initialized randomly.

4. Evaluate all individuals in the population based on the fitness function, *f*.

5. Select the best individuals based on *Step 4* and store them in a mating pool.

6. Perform the crossover process on the parents in the mating pool with a probability of *CXPB*.

7. Perform the mutation process on the offspring results from *Step 8* with a probability of *MUTPB*.

8. Generate a new population consisting of all the individuals from *Step 6*, *Step 7*, and the rest of the individuals from the current population.

9. Replace the current population with the new population.

10. Repeat *Steps 6* to *9 NGEN* times.

Now, let's look at a more concrete example of how a GA works in general. We will use the same objective function that we used in *Chapter 4, Evaluating Machine Learning Models* and treat this as a minimization problem. The objective function is defined as follows:

$$f(x) = \cos(6x) \cdot (1 - \sin(x^5)) + \varepsilon$$

Here, ε is the noise that follows the standard normal distribution. We are only going to perform a search within the $[-2,2]$ range. It is worth noting that in this example, we assume that we know what the true objective function is. However, in practice, this function is unknown. In this case, each individual will only have one gene, which is the value of x itself.

Let's say we define the hyperparameters for the GA method as *NPOP = 25, CXPB = 0.5, MUTPB = 0.15,* and *NGEN = 6*. As for the strategy of each **genetic operator**, we are using the **Tournament**, **Blend**, and **PolynomialBounded** strategies for selection, crossover, and mutation operators, respectively. The *Tournament* selection strategy works by selecting the best individuals among *tournsize* and the randomly chosen individual's *NPOP* times, where *tournsize* is the number of individuals participating in the tournament. The *Blend* crossover strategy works by performing a linear combination between two continuous individual genes, where the weight of the linear combination is governed by the *alpha* hyperparameter. The *PolynomialBounded* mutation strategy works by passing continuous individual genes to a predefined polynomial mapping.

There are many strategies available that you can follow based on your hyperparameter space specification. We will talk more about different strategies and how to implement the GA method using the **DEAP** package in *Chapter 10, Advanced Hyperparameter Tuning with DEAP and Microsoft NNI*. For now, let's see the results of applying a GA on the dummy objective function, *f*. Note that the points in each plot correspond to each individual in the population:

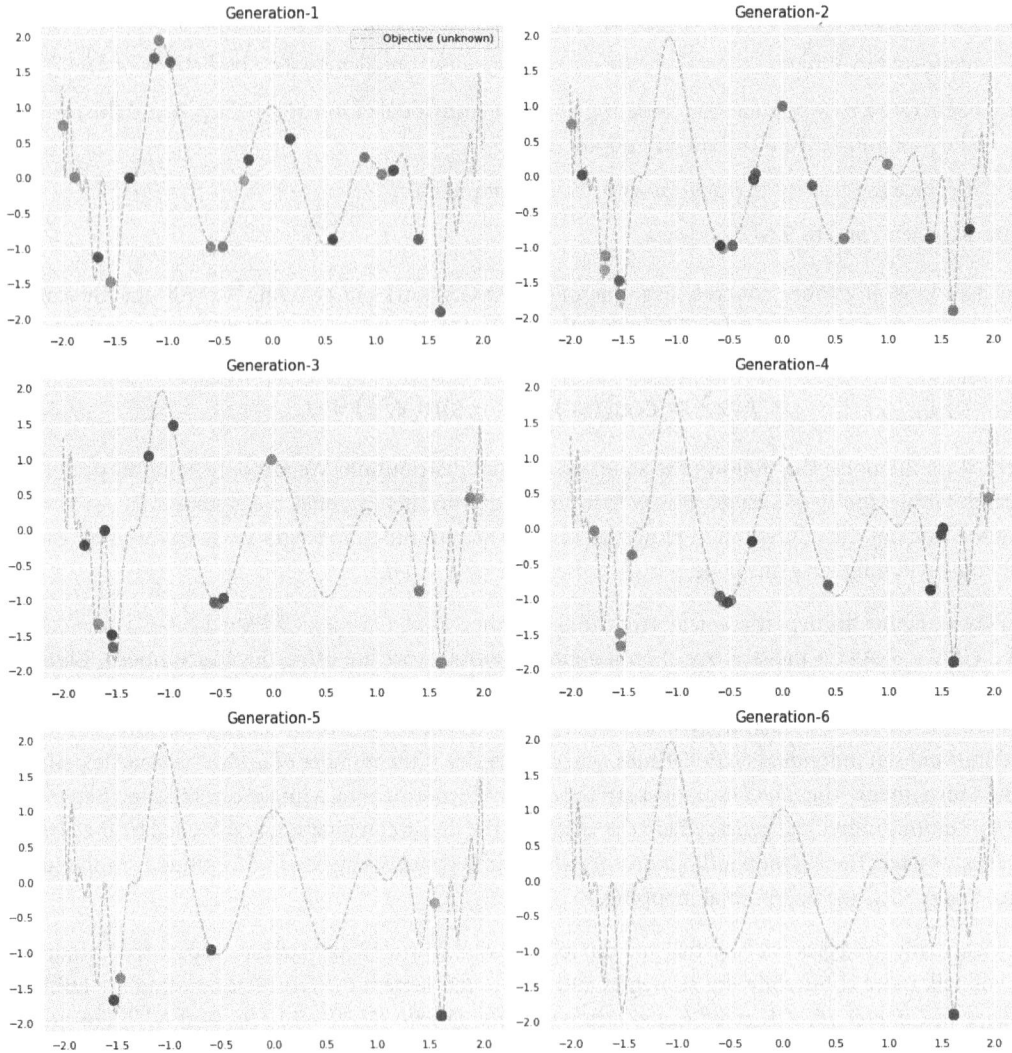

Figure 5.7 – GA process

Based on the preceding figure, we can see that in the first generation, individuals are scattered all around the place since it is initialized randomly. In the second generation, several individuals that are initialized around point −1.0 moved to other places that have lower fitness scores. However, in the third generation, there are new individuals around point −1.0 again. This may be due to the random mutation operator that's been applied to them. There are also several individuals stuck in the local optima, which is around point −0.5. In the fourth generation, most of the individuals have moved to places with lower fitness scores, although some of them are still stuck in the local optima. In the fifth generation, individuals are starting to converge in several places.

Finally, in the sixth generation, all of them converged to the near-global optima, which is around point 1.5. Note that we still have *NPOP=25* individuals in the sixth generation, but all of them are located in the same place, which is why you can only see one dot in the plot. This also applies to other generations if you see that there are fewer than 25 individuals in the plot. The convergence trend across generations can be seen in the following graph:

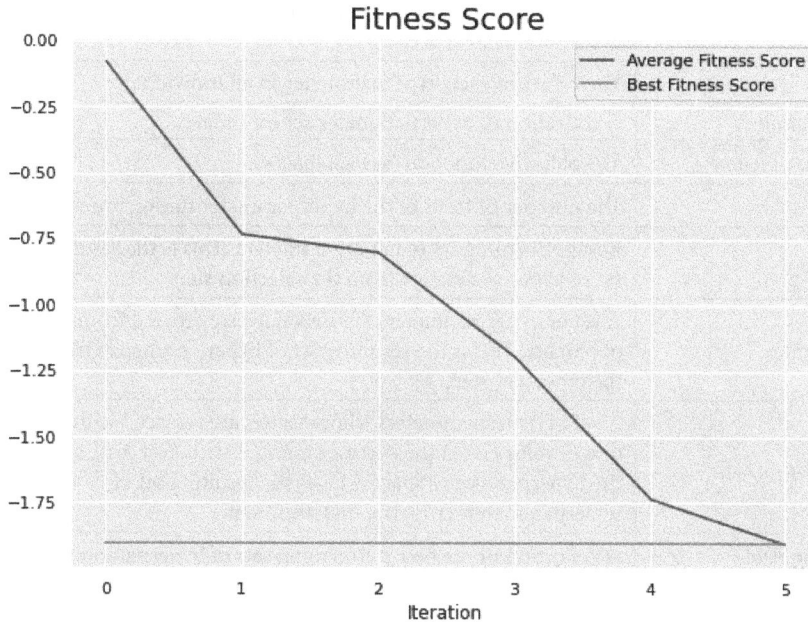

Figure 5.8 – Convergence plot

The trend that's shown in the preceding graph matches our previous analysis. However, we can get additional information from this plot. At first, many of the individuals are located in places with high fitness scores, but some individuals already get the best fitness score. Across generations, most of the individuals started to converge, and finally, in the last generation, all individuals had the best fitness score. It is worth noting that, in practice, it is not guaranteed that a GA will achieve the global optimal solution.

At this point, you may be wondering, how can a GA be adopted as a hyperparameter tuning method? What is the corresponding definition of all terms in the GA within the context of hyperparameter tuning? What does an individual mean when performing hyperparameter tuning with a GA?

As a hyperparameter tuning method, the GA method treats a set of hyperparameters as an individual where the hyperparameter values are the genes. To have better clarity on what each important term in the GA method means, in the context of hyperparameter tuning, please refer to the following table:

GA Terms	Definition in the Hyperparameter Tuning Context
Individual	A candidate set of hyperparameters that is sampled from the hyperparameter space.
Genes	The values of each hyperparameter in an individual.
Population	A collection of hyperparameter set candidates.
Fitness function	The objective function (see *Chapter 4*).
Generation	The number of trials of the hyperparameter tuning process.
Parent	Best performing set of hyperparameters. This is the resulting set of hyperparameters from the selection step.
Offspring	A set of hyperparameters whose values are crossed from pairs of parents. This is the resulting set of hyperparameters from the crossover step.
Mutant	A set of hyperparameters whose values are replaced with drawn values from the corresponding distribution within the hyperparameter space. This is the resulting set of hyperparameters from the mutation step.
Mating Pool	A collection of the best-performing sets of hyperparameters.

Figure 5.9 – Definition of GA method terms in the hyperparameter tuning context

Now that you are aware of the corresponding definition of each important term in the GA method, we can define the formal procedure to utilize *the GA method as a hyperparameter tuning method*:

1. Split the original full data into train and test sets.

2. Define the hyperparameter space, *H*, with the accompanied distributions.

3. Define the population size, *NPOP*.

4. Define the crossover probability, *CXPB*, and mutation probability, *MUTPB*.

5. Define the number of trials, *NGEN*, as the stopping criterion.

6. Define the objective function, *f*, based on the train set.

7. *Initialize* a population with *NPOP* sets of hyperparameters, where each set is drawn randomly from the hyperparameter space, *H*.

8. Evaluate all hyperparameter sets in the population based on the objective function, *f*.

9. *Select* several best candidate sets based on *Step 8*.

10. Perform *crossover* on candidate sets from *Step 9* with a probability of *CXPB*.

11. Perform *mutation* on the crossed candidate sets from *Step 10* with a probability of *MUTPB*.

12. Generate a new population consisting of all sets of hyperparameters from *Step 10*, *Step 11*, and the rest of the sets from the current population. The new population will also consist of *NPOP* sets of hyperparameters.

13. Repeat *Steps 8* to *12 NGEN* times.

14. Train on the full training set using the final hyperparameter values.

15. Evaluate the final trained model on the test set.

It is worth noting that when utilizing a GA as a hyperparameter tuning method, the GA itself has four hyperparameters, namely *NPOP*, *CXPB*, *MUTPB*, and *NGEN*, that control the performance of the hyperparameter tuning results, as well as the *exploration versus exploitation trade-off*. To be more precise, *CXPB* and *MUTPB*, or the crossover and mutation probability, respectively, are responsible for controlling the *exploration* rate, while the *selection* step, along with its strategy, controls the *exploitation* rate.

The following table lists the pros and cons of using a GA as a hyperparameter tuning method:

Pros	Cons
Works well with parallel computing resources.	Not suitable for expensive objective functions.
Exploitation is based on multiple candidates.	High computational cost due to the need to evaluate all individuals in each generation.
Works not only with numerical but also categorical types of hyperparameters.	

Figure 5.10 – Pros and Cons of the GA method

The need to evaluate all individuals in each generation means we multiplied the original time complexity that our objective has by *NPOP * NGEN*. It's very costly! That's why the GA method is not suitable for you if you have an expensive objective function and/or low computational resources. However, if you do have time to wait for the experiment to be done, and you have massively parallel computing resources, then the GA method is suitable for you. From a theoretical perspective, the GA method can also work with various types of hyperparameters – we just need to choose the appropriate crossover and mutation strategies for the corresponding hyperparameters. The GA method is better than SA in terms of having a population to guide which part of the subspace needs to be exploited more. However, it is worth noting that the GA method can still be stuck in local optima.

In this section, we discussed the GA method, starting with what it is, how it works both in terms of its general setup and the hyperparameter tuning context, and its pros and cons. We will discuss another interesting population-based heuristic search method in the next section.

Understanding particle swarm optimization

PSO is also a population-based heuristic search method, similar to the GA method. PSO is inspired by the schools of fish and flocks of birds' social interaction in nature. As a hyperparameter tuning method, PSO is suggested to be utilized if your search space contains many non-categorical hyperparameters, each trial doesn't take much time, and you have enough computational resources – especially parallel computing resources.

PSO is one of the most popular methods within the bigger **swarm intelligence (SI)** group of methods. There are various methods in SI that are inspired by the social interaction of animals in nature, such as herds of land animals, colonies of ants, flocks of birds, schools of fish, and many more. The common characteristics of SI methods are *population-based*, individuals within the population are relatively *similar to each other*, and the ability of the population to move in a specific direction systemically *without a single coordinator* inside or outside the population. In other words, the population can organize themselves based on the *local interactions* of individuals interacting with each other and/or the surrounding environment.

When a flock of birds is looking for food, it is believed that each bird can contribute to the group by sharing information about their sights, so that the group can move in the right direction. PSO is a method that simulates the movement of a flock of birds to optimize the objective function. In PSO, the flock of birds is called a **swarm** and each bird is called a **particle**.

Each particle is defined by its **position** vector and **velocity** vector. The movement of each particle consists of both stochastic and deterministic components. In other words, the movement of each particle is not only based on a predefined rule but is also influenced by random components. Each particle also remembers its own **best position**, which gives the best objective function value along the trajectory it has passed. Then, along with the **global best position**, it is used to update the velocity and position of each particle at a particular time. The global best position is just the position of the best particle from the previous step.

Let's say that $x_i(t)$ is the position vector in a d-dimensional space of the ith particle out of m particles in the swarm, and that $v_i(t)$ is the velocity vector of the same size for the ith particle, as shown here:

$$x_i(t) = [x_{i1}(t), \ x_{i2}(t), \ x_{i3}(t), \ ..., \ x_{id}(t)]^T$$

$$v_i(t) = [v_{i1}(t), \ v_{i2}(t), \ v_{i3}(t), \ ..., \ v_{id}(t)]^T$$

Let's also define the best position for each particle and the global best position vectors, respectively:

$$p_{bi}(t) = [p_{bi1}(t), p_{bi2}(t), p_{bi3}(t), \ldots, p_{bid}(t)]^T$$

$$g_b(t) = [g_{b1}(t), g_{b2}(t), g_{b3}(t), \ldots, g_{bd}(t)]^T$$

The following formulas define how each particle's position and velocity vectors are updated in each iteration:

$$x_i(t+1) = x_i(t) + v_i(t+1)$$

$$v_i(t+1) = \omega \cdot v_i(t) + c_1 \cdot r_1 \cdot \left(p_{bi}(t) - x_i(t)\right) + c_2 \cdot r_2 \cdot \left(g_b(t) - x_i(t)\right)$$

Here, ω, c_1, and c_2 are the hyperparameters that control the *exploration versus exploitation trade-off*. ω has a value between zero and one is usually called the **inertia weight coefficient**, while c_1 and c_2 are called the **cognitive** and **social coefficients**, respectively. r_1 and r_2 are the random values between zero and one and act as the stochastic components of the particle movement. Note that the d-dimensions of the position and velocity vectors refer to the number of hyperparameters we have in the search space, while the m particles refer to the number of candidate hyperparameters that are sampled from the hyperparameter space.

Updating the velocity vector may seem intimidating the first time, but actually, you can understand it more easily by treating the formula as three separate parts. The first part, or the left-most side of the formula, aims to update the next velocity proportional to the current velocity. The second part, or the middle part of the formula, aims to update the velocity toward the direction of the best position that the ith particle has, while also adding a stochastic component to it. The third part, or the right-most side of the formula, aims to bring the ith particle closer to the global best position, with additional random behavior applied to it. The following diagram helps illustrate this:

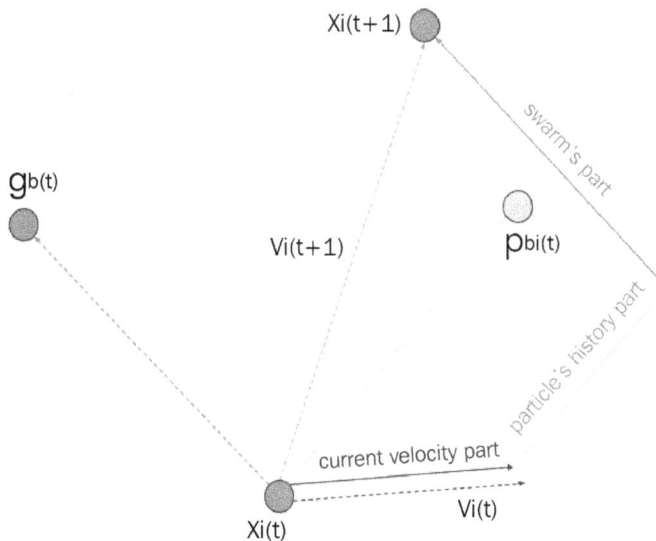

Figure 5.11 – Updating the particle's position and velocity

The preceding diagram isn't the same as the stated formula since the random components and the hyperparameters are missing from the picture. However, this diagram can help us understand the high-level concept of how each particle's position and velocity vectors are updated in each iteration. We can see that the final updated velocity (*see the orange line*) is calculated based on three vectors, namely the current velocity (*see the brown line*), the particle best position (*see the green line*), and the global best position (*see the purple line*). Based on the final updated velocity, we can get the updated position of the ith particle – that is, $x_i(t + 1)$.

Now, let's discuss how the hyperparameters affect the formula. The inertia weight coefficient, ω, controls how much we want to put our focus on the current velocity when updating the velocity vector. On the other hand, the cognitive coefficient, c_1, and the social coefficient, c_2, control how much we should focus on the particle's past trajectory history and swarm's search result, respectively. When we set $c_1 = 0$, we don't take into account the influence of the best position of the ith particle, which may lead us to be *trapped in the local optima*. When we set $c_2 = 0$, we ignore the influence of the global best position, which may lead us to a *slower convergence* speed.

Now that you are aware of the position and velocity components of each particle in the swarm, take a look at the following steps, which define *how PSO works in general* as an optimization method:

1. Define the swarm size, N, the inertia weight coefficient, w, the cognitive coefficient, c1, the social coefficient, c2, and the maximum number of trials.

2. Define the fitness function, f.

3. Initialize a swarm with N particles, where each particle's position and velocity vectors are initialized randomly.

4. Set each particle's current position vector as their best position vector, pbi.

5. Set the current global best position, gb, by selecting a position vector from all N particles that have the most optimal fitness score.

6. Update each particle's position and velocity vector based on the updating formula.

7. Evaluate all the particles in the swarm based on the fitness function, f.

8. Update each particle's best position vectors, pbi:

    I.   Compare each particle's current fitness score from *Step 7* with its *pbi* fitness score.

    II. If the current fitness score is better than the *pbi* fitness score, update *pbi* with the current position vector.

9. Update the global best position vector, gb:

    I.   Compare each particle's current fitness score from *Step 7* with the previous *gb* fitness score.

    II. If the current fitness score is better than the *gb* fitness score, update *gb* with the current position vector.

10. Update each particle's position and velocity vector based on the updating formula.

11. Repeat *Steps 7* to *10* until the maximum number of trials is reached.

12. Return the final global best position, *gb*.

It is worth noting that the definition of the optimal fitness score (or a better fitness score in the previously stated procedure) will depend on what type of optimization problem you are trying to solve. If it is a minimization problem, then a smaller fitness score is better. If it is a maximization problem, then it is the other way around.

To have even a better understanding of how PSO works, let's go through an example. Let's define the fitness function as follows:

$$f(x, y) = x^2 + \cos(6x) + y^2 + (1 - \sin(5y))$$

Here, x and y are only defined within the [0,2] range. The following *contour plot* shows what our objective function looks like. We will learn more about how to implement PSO using the **DEAP** package in *Chapter 10, Advanced Hyperparameter Tuning with DEAP and Microsoft NNI*:

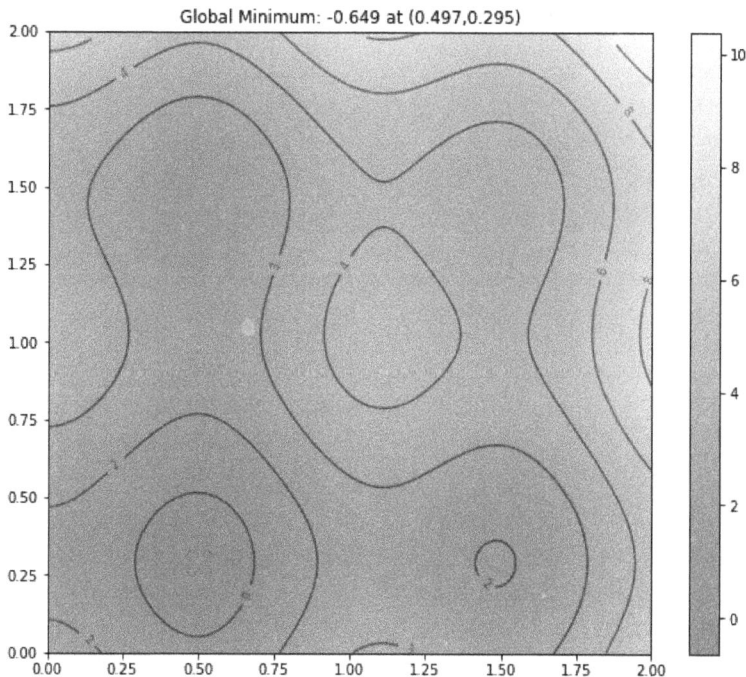

Figure 5.12 – A contour plot showing the objective function and its global minimum

Here, you can see that the global minimum (*see the red cross marker*) is located at (0.497, 0.295) with an objective function value of –0.649. Let's try to utilize PSO to see how well it estimates the minimum value of the objective function compared to the true global minimum. Let's say we define the hyperparameter for PSO as *N=20, w=0.5, c1=0.3*, and *c2=0.5*, and set the maximum number of trials to 16.

You can see the initial swarm illustration in the following contour plot. The blue dots refer to each of the particles, the blue arrow on each particle refers to the particle's velocity vector, the black dots refer to each particle's best position vectors, and the red star marker refers to the current global best position vector at a particular iteration:

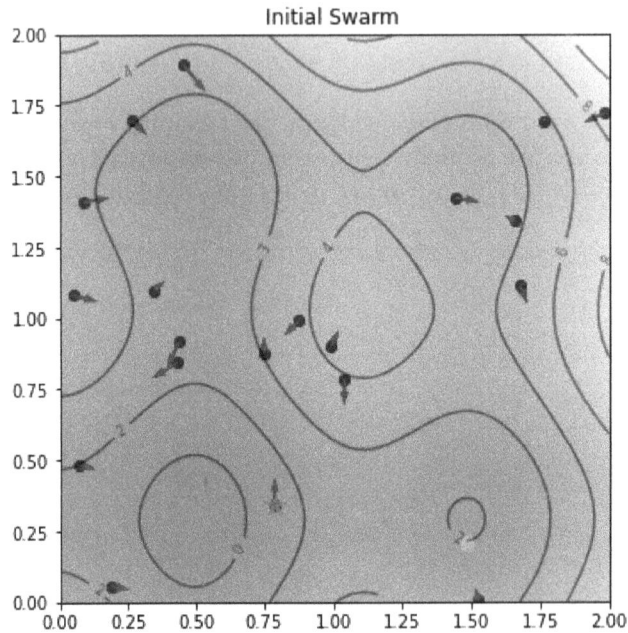

Figure 5.13 – A PSO initial swarm

Since the initial particles at the swarm are initialized randomly, the direction of the velocity vectors is all over the place (see *Figure 5.13*). You can see how each particle's position and velocity vectors are updated in each iteration, along with the global best position vector, as shown here:

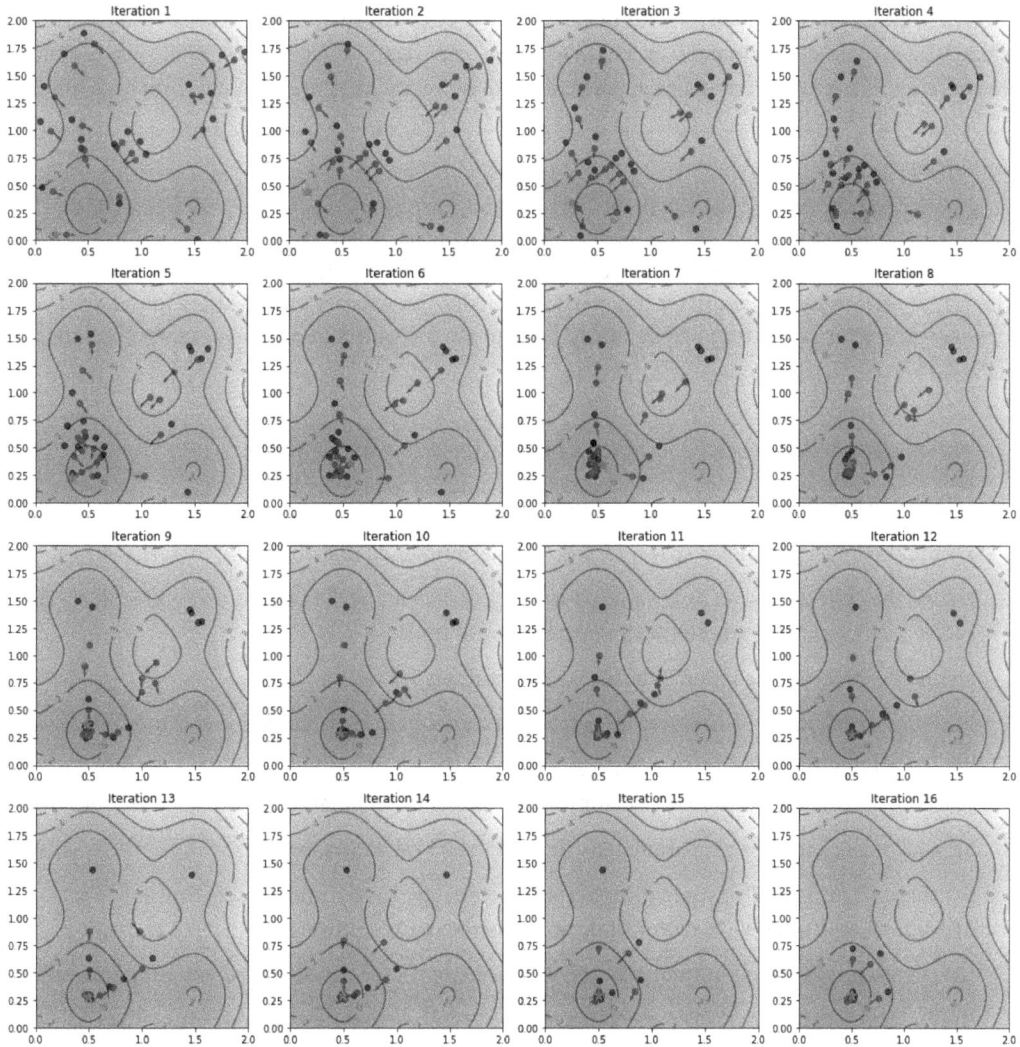

Figure 5.14 – PSO process

Even at the first iteration, each particle's velocity vector is pointing toward the global minimum, which is located in the bottom left of the plot. In each iteration, the position and velocity vectors are updated and move closer to the global minimum. At the end of the iteration loop, most of the particles are located around the global minimum position, where the final global best position vector is located at (0.496, 0.290) with a fitness score of around –0.648. This estimation is very close to the true global minimum of the objective function!

It is worth noting that the velocity vector of each particle contains two components: magnitude and direction. The magnitude will impact the length of the velocity vector in *Figure 5.14*. While you may not see the difference in length between each particle's velocity vector, they are different from each other!

> **Important Note**
>
> As a hyperparameter tuning method, in the PSO method, *particle* and *swarm* refer to the candidate set of hyperparameters that are sampled from the hyperparameter space and the collection of hyperparameter set candidates, respectively. The position vector of each particle refers to the values of each hyperparameter in a particle. Finally, the velocity vector refers to the *delta of hyperparameter values* that will be utilized to update the values of each hyperparameter in a particle.

The following steps define *how PSO works as a hyperparameter tuning method*:

1. Split the original full data into train and test sets.

2. Define the hyperparameter space, H, with the accompanied distributions.

3. Define the collection size, N, the inertia weight coefficient, w, the cognitive coefficient, $c1$, the social coefficient, $c2$, and the maximum number of trials.

4. Define the objective function, f, based on the train set.

5. Initialize a collection of N sets of hyperparameters, where each set is drawn randomly from the hyperparameter space, H.

6. Randomly initialize the velocity vector for each set of hyperparameters in the collection.

7. Set each set's current hyperparameter values as their best values, *pbi*.

8. Set the current global best set of hyperparameters, *gb*, by selecting a set from all N sets of hyperparameters that have the most optimal objective function score.

9. Update each set's hyperparameter values and velocity vector based on the updating formula.

10. Evaluate all sets of hyperparameters in the collection based on the objective function, f.

11. Update each set's best hyperparameter values, *pbi*:

    I.   Compare each set's current score from *Step 10* with its *pbi* score.

    II. If the current score is better than the *pbi* score, update *pbi* with the current hyperparameter values.

12. Update the global best set of hyperparameters, *gb*:

    I.   Compare each set's current score from *Step 10* with the previous *gb* score.

    II. If the current score is better than the *gb* score, update *gb* with the current set of hyperparameters.

13. Update each set's hyperparameter values and velocity vector based on the updating formula.

14. Repeat *Steps 10* to *13* until the maximum number of trials is reached.

15. Train on the full training set using the global best set of hyperparameters.

16. Evaluate the final trained model on the test set.

One issue with the updating formula in the PSO method is that it only works on numerical variables, especially continuous variables, meaning we can't directly utilize the original PSO as a hyperparameter tuning method if our hyperparameter space contains discrete hyperparameters. Motivated by this issue, there are several variants of PSO that are designed to be able to work in discrete spaces as well. The first variant is designed to work specifically for binary variables and is called **binary PSO**. In this variant, the updating formula for the velocity vector is the same, meaning we still treat the velocity vector in a continuous space, but the updating formula for the position vector is modified, like so:

$$x_{ij}(t+1) = 1 \; if \; r < sigmoid\left(v_{ij}(t+1)\right) \; else \; 0$$

Here, r is a random number drawn from a uniform distribution within the $[0,1)$, $sigmoid(z) = \frac{1}{1+e^{-z}}$ interval, and the j subscript refers to each component in the ith particle. As you can see, in the binary PSO variant, we can work within the discrete space, but we are restricted to only having binary variables.

What about when we have a combination of discrete and continuous numerical hyperparameters? For example, our hyperparameter space for a neural network model contains the learning rate, dropout rate, and the number of layers. We can't utilize the original PSO method directly since the number of layers hyperparameter expects an integer input, not a continuous or floating-point input. We also can't utilize the binary PSO variant since the learning rate and dropout rate are continuous, and the number of layers hyperparameter is also not binary.

One simple thing we can do is *round the updated velocity* vector component values, but only for components that correspond to the discrete position component, before passing it to the position vector updating formula. This way, we can ensure that our discrete hyperparameters will still always be within the discrete space. However, this workaround still has an issue. The rounding operation may make the updating procedure of the velocity vector suboptimal. Why? Because of the possibility that no matter the updated values of the velocity vector, so long as they are still within a similar range of one integer point, then the position vector will not be updated anymore. This will contribute to a lot of redundant computational costs.

There is another workaround to make PSO operate well both in continuous and discrete spaces. On top of rounding the updated velocity vector component values, we can also *update the inertia weight coefficient dynamically*. The motivation is to help a particle focus on its past velocity values so that it is not stuck in the local or global optimum, which is influenced by p_{bi} or g_b. The dynamic inertia weight updating procedure can be done based on several factors, such as the relative distance between its current position vector and its best position vector, the difference between the current number of trials and the maximum number of trials, and many more.

There are many variants of how we can dynamically update the inertia weight coefficient during trials; we will leave it to you to choose what works well for your specific case.

Although we can modify the updating formula in PSO to make it work not only for continuous but also discrete variables, we are still faced with several issues, as stated previously. Thus, to utilize the maximum power of PSO within the continuous space, there's another variant of PSO that tries to synergize PSO with the Bayesian optimization method, called **PSO-BO**. The goal of PSO-BO is to utilize PSO as a replacement for Bayesian optimization's acquisition function optimizer (see *Chapter 4*). So, rather than using a second-order optimization method to optimize the acquisition function, we can utilize PSO as the optimizer to help decide which set of hyperparameters to be tested in the next trial of the Bayesian optimization hyperparameter tuning procedure.

The following table summarizes the pros and cons of utilizing PSO as a hyperparameter tuning method:

Pros	Cons
Works well with parallel computing resources.	Not suitable for expensive objective functions.
Exploitation is done based on multiple candidates.	High computation cost due to the need to evaluate all individuals in each generation.
	Only works well with continuous types of hyperparameters, but can be modified for discrete hyperparameters as well.

Figure 5.15 – Pros and cons of PSO

Now that you are aware of what PSO is, how it works, its several variants, and its pros and cons, let's discuss another interesting population-based heuristic search method.

Understanding Population-Based Training

PBT is a population-based heuristic search method, just like the GA method and PSO. However, PBT is not a nature-inspired algorithm like GA or PSO. Instead, inspired by the GA method itself. PBT is suggested for when you are working with a neural-network-based type of model and *just need the final trained model* without knowing the specifically chosen hyperparameter configurations.

PBT is specifically designed to *work only with a neural network-based* type of models, such as a multilayer perceptron, deep reinforcement learning, transformers, GAN, and any other neural network-based models. It can be said that PBT does both hyperparameter tuning and *model training* since the weights of the neural network model are inherited during the process. So, PBT is not only for choosing the most optimal hyperparameter configurations but also for transferring the weights or parameters of the model to other individuals within the population. That's why the output of PBT is not a hyperparameter configuration but a model.

PBT is a *hybrid* method of the *random search* and *sequential search* methods, such as manual search and Bayesian search (see *Chapter 3, Exploring Exhaustive Search* and *Chapter 4, Exploring Bayesian Optimization* for more details). Random search is a very good method for finding a good subspace for sensitive hyperparameters. Sequential search methods tend to give better performance than random search if we have enough computational resources and time to execute the optimization process. However, the fact that those methods need to be executed sequentially makes the experiment take a very long time to run. PBT comes with a solution to combine the best of both worlds into a *single training optimization process*, meaning the model training and hyperparameter tuning process are merged into a single process.

The term *Population-Based* in PBT comes from the fact that it is inspired by the GA method in terms of utilizing knowledge of the whole population to produce a better-performing individual. Note that the **individual** part of PBT refers to each of the N models with different parameters and hyperparameters in the **population** or a collection of all those N models.

The search process in PBT starts by *initializing a population, P,* that contains N models, $\{M_i\}_{i=1}^N$, with their own randomly sampled parameters, $\{\theta_i\}_{i=1}^N$, and randomly sampled hyperparameters, $\{h_i\}_{i=1}^N$. Within each iteration of the search process, the *training step* is triggered for each of the N models. The training step consists of both forward and backward propagation procedures that utilize gradient-based optimization methods, just like the usual training procedure for a neural network-based model. Once the training step is done, the next step is to perform an *evaluation step*. The purpose of the evaluation step is to evaluate the current model's Mi performance on the unseen validation data.

Once the model, Mi, is considered *ready*, PBT will trigger the *exploit* and *explore* steps. The definition of a model being ready may vary, but we can define "ready" as passing a predefined number of steps or passing a predefined performance threshold. Both the exploit and explore steps have the same goal, which is to update the model's parameters and hyperparameters. The difference is determined by how they do the update process.

The **exploit** step will decide, based on the evaluation results from the whole population, whether to keep utilizing the current set of parameters and hyperparameters or to focus on a more promising set. For example, the exploit step can be done by replacing a model that is considered as part of the bottom X% models in the whole population with a randomly sampled model from the top X% models in the population. Note that a model consists of all the parameters and hyperparameters. On the other hand, the **explore** step updates the model's set of hyperparameters, *not parameters*, by proposing a new set. You can propose a new set by randomly perturbing the current set of hyperparameters with a predefined probability or by resampling the set of hyperparameters from the top X% models in the population. Note that this exploration step is only done on the chosen model from the exploitation step.

> **Important Note**
>
> The exploration step in PBT is inspired by random search. This step can identify which subspace of hyperparameters needs to be explored more using *partially trained models* chosen from the exploitation step. The evaluation step that is done within the search process also enables us to remove the drawback of the sequential optimization process.

The exploitation and exploration procedure in the PBT method allows us to update a model's set of hyperparameters in an *online fashion*, while also putting more focus on the promising hyperparameter and weight space. The iterative process of train-eval-exploit-explore is performed *asynchronously in parallel* for each of the N individuals in the population until the stopping criterion is met.

The following steps summarize *how PBT works as a single training optimization process*:

1. Split the original full data into train, validation, and test sets (see *Chapter 1, Evaluating Machine Learning Models*).

2. Define the hyperparameter space, H, with the accompanied distributions.

3. Define the population size, N, the exploration perturbation factor, *perturb_fact*, the exploration resampling probability, *resample_prob*, and the exploitation fraction, *frac*.

4. Define the model's *readiness criterion*. Usually, the number of SGD optimization steps is used. However, it is also possible to use the model's performance threshold as the criterion.

5. Define the *checkpoint directory* that is used to store the model's weights and hyperparameters.

6. Define the evaluation function, f.

7. Initialize a population, P, that contains N models, $\{M_i\}_{i=1}^{N}$, with their own randomly sampled parameters, $\{\theta_i\}_{i=1}^{N}$, and randomly sampled hyperparameters, $\{h_i\}_{i=1}^{N}$, from the hyperparameter space, H.

8. For each model in the population, P, run the following steps *in parallel*:

    I.   Run one step of the training process for the model, M_i, with the θ_i parameter and a set of hyperparameters, h_i.

    II.   If the *readiness criterion* has been met, do the following. If not, go back to *Step I*:

    -   Perform the *evaluation* step based on f on the validation set.

    -   Perform the *exploitation* step on the model, M_i, based on the predefined exploitation fraction, *frac*. This step will result in a new set of parameters and hyperparameters.

    -   Perform the *exploration* step on the set of hyperparameters from the exploitation step based on the predefined *perturb_fact* and *resample_prob*.

    -   Perform the *evaluation* step on the new set of parameters and hyperparameters based on f on the validation set.

    -   *Update* the model, M_i, with the new set of parameters and hyperparameters.

    III. Repeat Steps I and II until the end of the training loop. Usually, it is defined by the number of epochs.

9. Return the model with the best evaluation score in the population, *P*.

10. Evaluate the final model on the test set.

It is worth noting that, in practice, such as in the implementation of the **NNI** package (see *Chapter 10, Advanced Hyperparameter Tuning with DEAP and Microsoft NNI*), the readiness criterion defined in *Step 4* is an epoch. In other words, the second step within *Step 8* will only be run after each training epoch, not in the middle of an epoch. It is also worth noting that the checkpoint directory defined in *Step 5* is needed because, in PBT, we need to copy weights from another model in the population, while that's not the case for the other hyperparameter tuning methods we've learned about so far.

While the original PBT algorithm states that we can run *Step 8* asynchronously in parallel, this is not the case in the implementation of the **NNI** package, which will be used in this book to implement PBT. In the NNI package implementation, the process is run *synchronously*, meaning that we can continue to the next epoch once all of the individuals or models in the population have finished the previous epoch.

The following table lists the pros and cons of the PBT method:

Pros	Cons
Acts not only as a hyperparameter tuning method but also as a model training algorithm.	Works only on neural network-based models.
Works well with parallel computing resources.	High computation cost due to the need to evaluate all individuals in each iteration.
Exploitation is done based on multiple candidates.	

Figure 5.16 – Pros and cons of PBT

In this section, you learned all you need to know about PBT, including what it is, how it works, what makes it different from other heuristic search methods, and its pros and cons.

Summary

In this chapter, we discussed the third out of four groups of hyperparameter tuning methods, called the heuristic search group. We discussed what the heuristic search method is in general and several variants of heuristic search methods, including SA, the GA method, PSO, and PBT. We saw what makes each of the variants differ from each other, along with the pros and cons of each. At this point, you should be able to explain heuristic search in confidence when someone asks you. You should also be able to debug and set up the most suitable configuration of the chosen method that suits your specific problem definition.

In the next chapter, we will start discussing multi-fidelity optimization, the last group of hyperparameter tuning methods. The goal of the next chapter is similar to this one's: to provide a better understanding of the methods that belong to the multi-fidelity optimization group so that you can explain those methods in confidence when someone asks you. By doing this, you will be able to configure each of the methods for your specific problem!

6

Exploring Multi-Fidelity Optimization

Multi-Fidelity Optimization (**MFO**) is the fourth of four groups of hyperparameter tuning methods. The main characteristic of this group is that all methods belonging to this group utilize the cheap approximation of the whole hyperparameter tuning pipeline so we can have similar performance results with a much lower computational cost and faster experiment time. This group is suitable when you have a very large model or a very large number of samples, for example, when you are developing a neural-network-based model.

In this chapter, we will discuss several methods in the MFO group, including coarse-to-fine search, successive halving, hyper band, and **Bayesian Optimization and Hyperband** (**BOHB**). As in *Chapter 5, Exploring Heuristic Search* we will discuss the definition of each method, the differences between them, how they work, and the pros and cons of each.

By the end of this chapter, you will be confident in explaining MFO and its variations, and also how they work at a high level and in a technical way. You will also be able to tell the differences between them, along with the pros and cons of each. You will also experience the crucial benefit of understanding each of the methods in practice: being able to configure the method to match your own problem and knowing what to do when there are errors or unexpected outputs from the method.

In this chapter, we'll be covering the following main topics:

- Introducing MFO
- Understanding coarse-to-fine search
- Understanding successive halving
- Understanding hyper band
- Understanding BOHB

Introducing MFO

MFO is a group of hyperparameter tuning methods that work by creating a cheap approximation of the whole hyperparameter tuning pipeline so that we can get similar performance results with much *lower computational cost* and *faster experiment time*. There are many ways to create a cheap approximation. For example, we can work only on the subsets of the full data in the first several steps rather than directly working on the full data, or we can also try to use fewer epochs when training a neural-network-based model before training our model with full epochs. In other words, MFO methods work by *combining cheap low-fidelity and expensive high-fidelity* evaluations, where usually the proportion of cheaper evaluations is much larger than the more expensive evaluations so that we can achieve lower computational cost and thus faster experiment time. However, MFO methods can also be categorized as part of the **informed search** category since they utilize knowledge from previous iterations to have a (hopefully) better search space in future.

All of the methods that we have learned in the previous chapters can be categorized as **black-box optimization** methods. All black-box optimization methods try to perform hyperparameter tuning without utilizing any information from what is happening inside the ML model or the data that is used by the model. A black-box optimizer will only focus on searching the best set of hyperparameters from the defined hyperparameter space and *treat other factors as a black box* (see *Figure 6.1*). This characteristic has its own good and bad implications. It enables us to utilize a black-box optimizer, which is more flexible for various types of models or data, but it also costs us more since we do not consider other factors that may speed up the process.

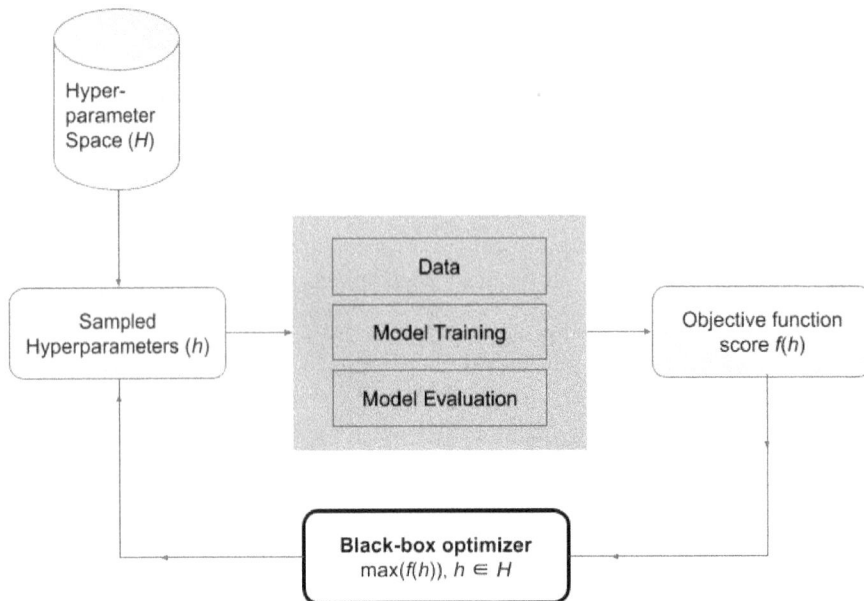

Figure 6.1 – Illustration of black-box optimizer

The expense of black-box optimization methods means we can't utilize them when we are working with a *very large model* or *big data* that requires a very long time for just one training iteration. That's where the MFO group of hyperparameter tuning methods comes into the picture! By considering other factors that are treated as black-box by black-box optimizers, we can have a faster process while sacrificing a bit of the generality that black-box optimizers have.

Generality

Generality means the model is able to perform on many unseen cases.

Furthermore, most of the methods categorized in this group can *utilize parallel computational resources* very nicely, which can further boost the speed of the hyperparameter tuning process. However, the benefit of faster processes offered by MFO methods comes with a cost. We may have *worse performing tuning results* since there is a chance we have excluded a better subspace during the cheap low-fidelity evaluations step. However, the *speedup is arguably more significant* than the estimation error, especially when we are working with a very large model and/or big data.

Important Note

The MFO group of hyperparameter tuning methods is *not* a completely different group compared to black-box optimization methods, including exhaustive search, Bayesian optimization, and heuristic search. In fact, we can also apply a similar procedure done in a multi-fidelity optimization method to a black-box optimizer. In other words, *we can combine black-box-and multi-fidelity* models so we can get the best of both worlds.

For example, we can perform hyperparameter tuning with one of the **Bayesian Optimization (BO)** methods (see *Chapter 4, Exploring Bayesian Optimization*) and also apply the successive halving method (see the *Understanding successive halving* section) on top of it. This way, we will ensure that we only perform BO on important subspace, rather than letting BO explore the whole hyperparameter space by itself. By doing this, we can have a faster experiment time with lower computational cost.

Now that you are aware of what MFO is, how it differs from black-box optimization methods, and how it works at a high level, we will dive deeper into several MFO methods in the following sections.

Understanding coarse-to-fine search

Coarse-to-Fine Search (CFS) is a combination of grid and random search hyperparameter tuning methods (see *Chapter 3, Exploring Exhaustive Search*). Unlike grid and random search, which are categorized in the **uninformed search** group of methods, CFS utilizes knowledge from previous iterations to have a (hopefully) better search space in the future. In other words, CFS is a *combination of sequential and parallel* hyperparameter tuning methods. It is indeed a very simple method since it is basically a *combination of two other simple methods: grid and random search*.

CFS can be effectively utilized as a hyperparameter tuning method when you are working with a medium-sized model, for example, a shallow neural network (other types of models can also work) and a moderate amount of training data.

The main idea of CFS is just to start with a *coarse* random search from the whole hyperparameter space, then gradually *refine* the search in more detail, either using random or grid search. The following figure summarizes how CFS works as a hyperparameter tuning method.

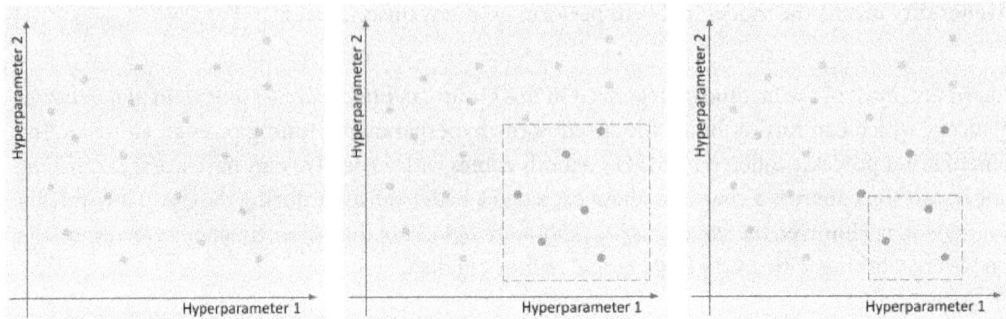

Figure 6.2 – Illustration of CFS

As illustrated in *Figure 6.2*, CFS starts by performing a random search in the whole pre-defined hyperparameter space. Then, it looks for a promising subspace based on the first coarse random search evaluation results. The definition of a promising subspace may vary and can be adjusted to your own preference. The following list shows several definitions of a promising subspace that you can adopt:

- Get only the *top N percentiles* of the best set of hyperparameters based on the evaluation performed in the previous trial.

- Put a *hard threshold* to filter out the bad set of hyperparameters from the previous trial.

- Conduct a *univariate analysis* to get the best range of values for each hyperparameter.

No matter what definition you are using to define the promising subspace, we will always get a list of values for each hyperparameter. Then, we can create a new hyperparameter space based on the minimum and maximum values in each list of hyperparameter values.

After getting the promising subspace, we can continue the process by performing a grid search or another random search in the smaller area. Note that you can also put a condition on when to keep using random search and when to start using grid search. Again, it is up to you to choose the appropriate condition. However, it is better to perform a random search than a grid search, so that we can have *more evaluations based on the cheap low-fidelity approach* compared to the expensive high-fidelity approach. We keep repeating this procedure until we reach the stopping criterion.

The following procedure explains in more detail how CFS works as a hyperparameter tuning method:

1. Split the original full data into a training set and a test set. (See *Chapter 1, Evaluating Machine Learning Models.*)

2. Define the hyperparameter space, H, with the accompanied distributions, the objective function, f, based on the training set, and the stopping criterion.

3. Define the grid size for creating the grid search hyperparameter space, `grid_size`, and the random search number of iterations, `random_iters`.

4. Define the criterion of a promising subspace by utilizing the objective function, f.

5. Define the criterion of when to start using grid search.

6. Set the initial best set of hyperparameters, `best_set`, with the value None.

7. Perform a random search on the current hyperparameter space, H, for `random_iters` times.

8. Select a promising subspace based on the criterion defined in *step 4*:

    I.   If the current best-performing set of hyperparameters is worse than the previous `best_set`, add `best_set` to the promising subspace.

    II.   If the current best-performing set of hyperparameters is better than the previous `best_set`, update `best_set`.

9. If the criterion in `step 5` is met, do the following:

    I.   Update the current hyperparameter space, H, with the promising subspace selected in *step 8*, using unique `grid_size` values for each of the hyperparameters.

    II.   Perform a grid search on the updated hyperparameter space, H.

10. If the criterion in *step 5* is not met, do the following:

    I.   Update the current hyperparameter space, H, with the promising subspace selected in *step 8* using the minimum and maximum values for each hyperparameter.

    II.   Perform a random search on the updated hyperparameter space, H, for `random_iters` times.

11. Repeat *steps 8 – 10* until the stopping criterion is met.

12. Train on the full training set using the best hyperparameter combination.

13. Evaluate the final trained model on the test set.

In CFS, the multi-fidelity characteristic is based neither on the amount of data nor the number of training epochs, but on the *granularity of the search* performed in the search space during each trial. In other words, we will *keep using all of the data* and *all of the training epochs* with a *refined hyperparameter space* in each trial.

Let's see how CFS works as a hyperparameter tuning method on dummy data generated by the **scikit-learn** package. Scikit-learn has a function called `make_classification` to create dummy classification data with several customizable configurations. In this example, we use the following configurations to generate the dummy data:

- *Number of classes.* We set the number of target classes in the data to 2 by setting `n_classes=2`.

- *Number of samples.* We set the number of samples to 500 by setting `n_samples=500`.

- *Number of features.* We set the number of features or the number of dependent variables in the data to 25 by setting `n_features=25`.

- *Number of informative features.* We set the number of features that have high importance to distinguish between all of the target classes to 18 by setting `n_informative=18`.

- *Number of redundant features.* We set the number of features that are basically just a weighted sum from other features to 5 by setting `n_redundant=5`.

- *Random seed.* To ensure reproducibility, we set `random_state=0`.

We utilize a **Multi-Layer Perceptron** (**MLP**) with one hidden layer as the classifier model and use the *mean of seven-fold cross-validation accuracy scores* as the objective function (see *Chapter 4, Exploring Bayesian Optimization*). In this example, we are not using grid search as part of the CFS procedure, meaning that we only use random search in each of the trials. We set the maximum number of trials to `12`, which acts as the stopping criterion. We set the number of iterations for each random search trial to `20`. Finally, we utilize the *top N percentiles* scheme to define the promising subspace in each trial, with N=`50`. We define the hyperparameter space as follows:

- Number of neurons in the hidden layer: `hidden_layer_sizes=range(1,51)`

- Initial learning rate: `learning_rate_init=np.linspace(0.001,0.1,50)`

The following figure shows how CFS works in each iteration or trial. The *purple dots* refer to hyperparameter values tested in the current trial, while the *red rectangles* refer to the promising subspace to be searched in the next trial.

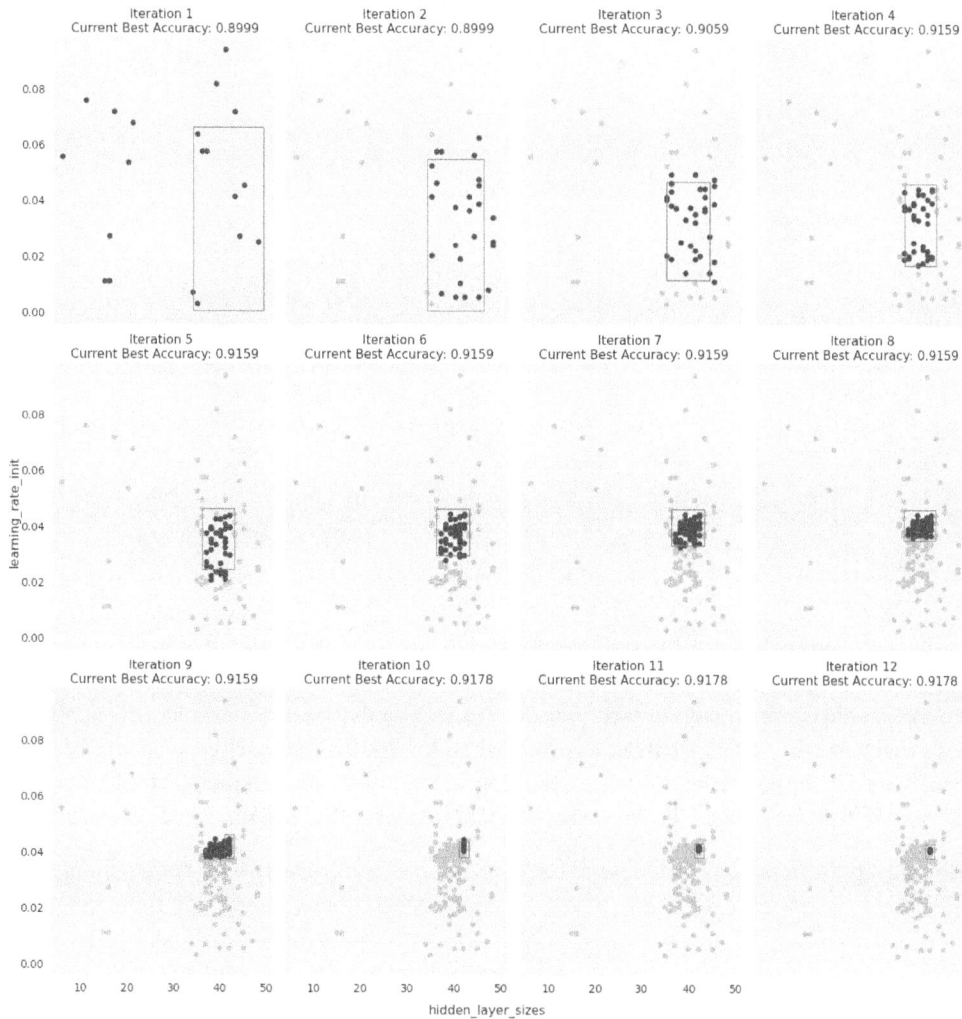

Figure 6.3 – Illustration of the CFS process

In *Figure 6.3*, we can see clearly how CFS starts by working at the full hyperparameter space and then gradually searches in the smaller subspaces. It is also worth noting that although we only use random search in this example, we can see that CFS still increases its fidelity over the number of trials until we get a final set of hyperparameters in the last trial. We can also see the performance of each trial in the following figure.

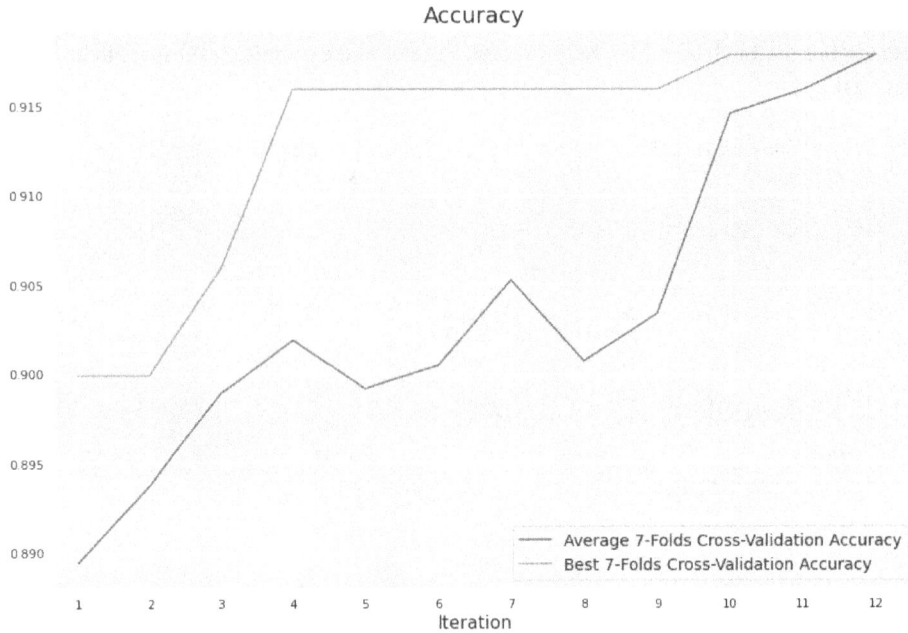

Figure 6.4 – Convergence plot

The blue line in *Figure 6.4* reflects the average cross-validation scores from all tested hyperparameters (see the purple dots in *Figure 6.3*) at each trial. The red line reflects the cross-validation score of the best-performing set of hyperparameters at each trial. We can see that the red line has a nice *non-decreasing monotonic* characteristic. This happens because we always add back the best set of hyperparameters from all previous trials to the promising subspace definition, as defined in *step 8* in the previous procedure. We will learn how to implement CFS with scikit-learn in *Chapter 7, Hyperparameter Tuning via Scikit*.

The following table summarizes the pros and cons of utilizing CFS as a hyperparameter tuning method.

Pros	Cons
Able to focus on more promising hyperparameter subspaces.	Utilizes all the data in each iteration, which makes it hard to apply when you have an enormous amount of data.
A very simple method with a customizable module.	No clear definition of how to choose a promising subspace.
	Simple but tricky to implement.

Figure 6.5 – Pros and cons of CFS

In this section, we have discussed CFS, looking at what it is, how it works, and the pros and cons. We will discuss another interesting MFO method in the next section.

Understanding successive halving

Successive Halving (SH) is an MFO method that is not only able to focus on a more promising hyperparameter subspace but can also *allocate computational cost wisely* in each trial. Unlike CFS, which utilizes all of the data in each trial, SH can utilize less data for a not-too-promising subspace while utilizing more data for a more promising subspace. It can be said that SH is a variant of CFS with a much clearer algorithm definition and is wiser in spending the computational cost. The most effective way to utilize SH as a hyperparameter tuning method is when you are working with a large model (for example, a deep neural network) and/or working with a large amount of data.

Similar to CFS, SH also *utilizes grid search or random search* to search for the best set of hyperparameters. At the first iteration, SH will perform a grid or random search on the whole hyperparameter space with a small amount of **budget** or resources, and then it will gradually increase the budget while also removing the worst half of the hyperparameters candidates at each iteration. In other words, SH performs hyperparameter tuning with a lower budget on a bigger search space and a higher budget on a more promising smaller subspace. SH can also be seen as a **tournament** between hyperparameter candidates, where only the best candidate will survive at the end of the trials.

> **Budget Definition in SH**
>
> In a default hyperparameter tuning setup, the budget is defined as the number of samples in the data. However, it is also possible to define the budget in other ways. For example, we can also define the budget as the maximum training time, number of iterations during XGBoost training steps, number of estimators in a random forest, or number of epochs when training a neural network model.

To have a better understanding of SH, let's look at the following example before we discuss how it works in a formal procedure. We utilize the same model and the same hyperparameter space definition used in the example in the *Understanding CFS* section. We also utilize a similar procedure to generate a dummy classification dataset a hundred times bigger in size, meaning we have 50000 samples instead of only 500 samples as in the CFS example.

In this example, we utilize random search instead of grid search to sample the hyperparameter candidates in each trial. The following figure shows the accuracy scores of hyperparameter candidates over trials. Each line refers to the trend of each hyperparameter candidate's objective function score, which in this case is the *seven-fold cross-validation accuracy score*, over the number of trials. The final objective function score, based on the best set of hyperparameters selected from the SH tuning process, is 0.984. We will learn how to implement SH in *Chapter 7, Hyperparameter Tuning via Scikit* and *Chapter 9, Hyperparameter Tuning via Optuna*.

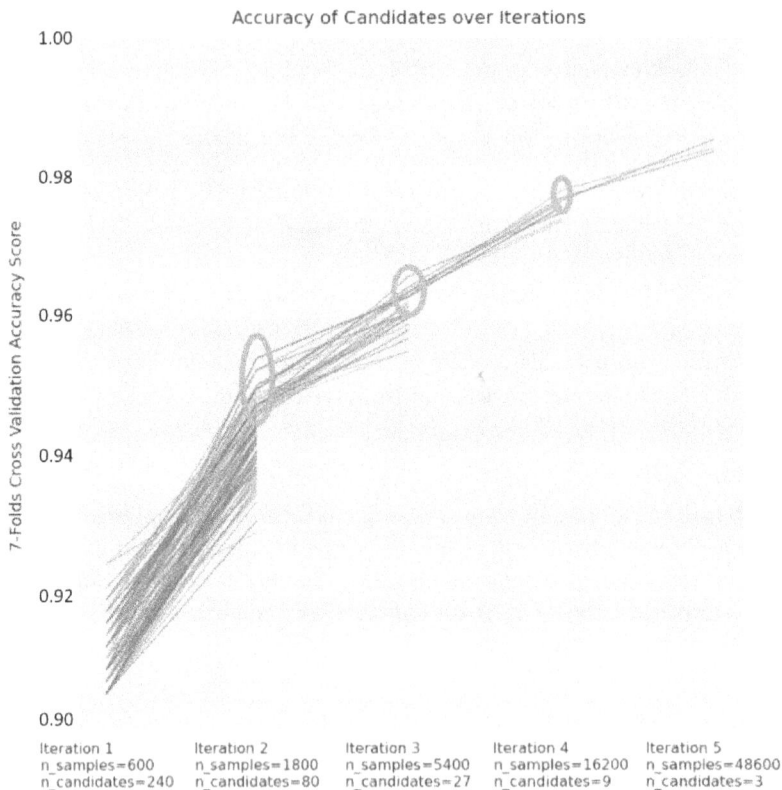

Figure 6.6 – Illustration of the SH process

In *Figure 6.6*, we can clearly see how SH takes only the top hyperparameter candidates (see the orange ovals) from each trial for further evaluation in the next trial. In the first iteration, a random search is performed 240 times with only 600 out of 50000 of the samples available in the data. This means we have 240 hyperparameter candidates, n_candidates, in the first iteration. Out of those hyperparameter candidates, SF takes only the top 80 candidates to be evaluated with a larger number of samples in the second iteration, which is 1800 samples. For the third iteration, SF again takes only the top 27 candidates and evaluates them on 5400 samples.

This process continues until we *can't use a larger number of samples* since it will be greater than the maximum resources, max_resources, defined in the first place. In this example, the maximum resources are defined as the number of samples that we have in the data. However, it can also be defined as the total number of epochs or training steps based on the definition of the budget or resources.

In this example, we stopped at the fourth iteration, where we need to evaluate 3 candidates based on 48600 samples. The final hyperparameter candidate chosen is the one that has the highest seven-fold cross-validation accuracy score evaluated on those 48600 samples.

As you will notice, the gradual increment of the number of samples in each trial and the gradual decrement of the number of candidates in each trial follows the same multiplier factor, factor, which is 3 in this example. That's why we have to stop at the fourth iteration, since if we continue to the fifth iteration, we would need 48600*3=145800 samples, while we only have 50000 samples in the data. Note that we have to set the value of the multiplier factor ourselves before running the SH tuning process. In other words, this multiplier factor is the hyperparameter for SH.

Multiplier Factor in SH

The halving term in SH refers to setting the multiplier factor value to two. In other words, only the best half of the hyperparameter candidates in each trial are passed to the next trial. However, we can also change this with another value. For example, when we set the multiplier factor as three, it means we take only the top one-third of hyperparameter candidates in each trial. In practice, setting the multiplier factor as three usually works better than setting it as two.

Besides the multiplier factor and maximum resources, SH also has other hyperparameters, such as the minimum number of resources to be used at the first iteration, min_resources, and the initial number of candidates to be evaluated at the first iteration, n_candidates. If grid search is utilized in the SH tuning process, n_candidates will equal the number of all combinations of hyperparameters in the search space. If a random search is utilized, then we have to set the value of n_candidates ourselves. In our example, where random search is utilized, we set min_resources=600 and n_candidates=240.

While setting factor to be equal to three is the common practice, this is not the case for min_resources and n_candidates. There are many factors to be considered before choosing the right values for both the min_resources and n_candidates hyperparameters. In other words, there is a trade-off between them, as explained here:

- Choosing a *bigger* value for n_candidates is useful when the bad and good hyperparameters can be easily distinguished with a smaller number of samples (a smaller value for min_resources).

- Choosing a *smaller* value for n_candidates is useful when we need a larger number of samples (a larger value for min_resources) to distinguish between the bad and good hyperparameters.

Another hyperparameter that SH has is the minimum early stopping rate, `min_early_stopping`. This integer-type hyperparameter has a default value of zero. If it is set to more than zero, it will reduce the number of iterations while increasing the number of resources to be used at the first iteration. In our previous example, we set `min_early_stopping=0`.

To summarize, SH as a hyperparameter tuning method works as follows:

1. Split the original dataset into train and test sets.

2. Define the hyperparameter space, H, with the accompanied distributions, and the objective function, f, based on the training set.

3. Define the budget/resources. Usually, this is defined as the number of samples or training epochs.

4. Define the maximum amount of resources, `max_resources`. Usually, this is defined as the total number of samples in data or the total number of epochs.

5. Define the multiplier factor, `factor`, the minimum amount of resources to be used at the first iteration, `min_resources`, and the minimum early stopping rate, `min_early_stopping`.

6. Define the initial number of hyperparameter candidates to be evaluated at the first iteration, `n_candidates`. If grid search is utilized, this will be automatically parsed from the total number of hyperparameter combinations in the search space.

7. Calculate the maximum number of iterations, n_{iter}, using the following formula:

$$n_{iter} = 1 + floor\left(\log_{factor}\left(\frac{max\_resources}{min\_resources}\right)\right)$$

8. Assert if `n_candidates` $\geq factor^{(n_{iter}-min\_earlystopping-1)}$ to ensure there is at least one candidate in the last iteration.

9. Warm up the first iteration:

   I. Sample `n_candidates` sets of hyperparameters from the hyperparameter space. If grid search is utilized, just return all of the hyperparameter combinations in the space. This set of candidates is referred to as *candidates1*.

   II. Evaluate all *candidates1* sets of hyperparameters, using `min_resources`, based on the objective function, *f*.

   III. Calculate the *topK* value that will be used to select top candidates for the next iteration:

   $$top_K = floor(n\_candidates \cdot factor^{-1})$$

10. For each iteration, *i*, starting from the second iteration until n_{iter} iteration, proceed as follows:

    I. Update the current set of candidates, *candidatesi*, by selecting *topK* candidates from *candidatesi-1* in terms of the most optimal objective function score.

II.   Update the current allocated resources, *resourcesi*, based on the following formula:

$$resources_i = min\_resources \cdot factor^{(i+min\_early\_stopping-1)}$$

III.   Evaluate all *candidatesi* sets of hyperparameters, using *resourcesi*, based on the objective function, *f*.

IV.   Update the *topK* value based on the following formula:

$$top_K = floor(n\_candidates \cdot factor^{-i})$$

11.   Return the best hyperparameter candidate:

I.   Evaluate all candidates in the last iteration using the allocated number of resources and the objective function, `f`. Note that it's possible that the allocated resource in the last iteration is less than `max_resources`.

II.   Select the candidate with the optimal objective function score.

12.   Train on the full train set using the best set of hyperparameters from *step 11*.

13.   Evaluate the final trained model on the test set.

Based on the previous example and the stated procedure, we can see that SH performs cheap, low-fidelity evaluations on the first several iterations by using a low number of resources and starts to perform more expensive high-fidelity evaluations on the final several iterations by using a high number of resources.

Integration with Other Black-Box Methods

SH can also be utilized along with other black-box hyperparameter tuning methods apart from grid and random search. For example, in the **Optuna** (see *Chapter 9, Hyperparameter Tuning via Optuna*) package, we can combine TPE (see *Chapter 4, Exploring Bayesian Optimization*) with SH, where SH acts as a **pruner**. Note that in Optuna, the budget/resources is defined as the number of training steps or epochs instead of the number of samples.

The following is a list of the pros and cons of SH as a hyperparameter tuning method:

Pros	Cons
Able to focus on a more promising hyperparameter subspace.	Hard to find the sweet spot as regards the trade-off between the amount of resources and the number of candidates.
Able to allocate computational cost wisely in each trial.	A better set of hyperparameters is possibly removed during the first several iterations.

Figure 6.7 – Pros and cons of SH

In practice, most of the time, we do not know how to balance the trade-off between the number of resources and the number of candidates since there is no clear definition of how to distinguish bad and good hyperparameters. One thing that can help us to find a sweet spot in this trade-off is leveraging previous similar experiment configurations or by performing **meta-learning** based on the available meta-data from previous similar experiments.

Now you are aware of SH, how it works, when to use it, and its pros and cons, in the next section, we will learn about an extension of this method that attempts to overcome the cons of SH.

Understanding hyper band

Hyper Band (HB) is an extension of SH that is specifically designed to overcome issues inherent in SH (see *Figure 6.7*). Although we can perform meta-learning to help us balance the trade-off, most of the time we do not have the metadata that's needed in practice. Furthermore, the possibility of SH removing better sets of hyperparameters in the first several iterations is also worrying and can't be solved by just finding a sweet spot from the trade-off. HB tries to solve these issues by calling SH several times iteratively.

Since HB is just an extension of SH, it is suggested that you utilize HB as your hyperparameter tuning method when you are working with a large model (for example, a deep neural network) and/or working with a large amount of data, just like SH. Furthermore, it is even better to utilize HB than SH when you do not have the time or metadata needed to help you configure the trade-off between the amount of resources and the number of candidates, which is the case most of the time.

The main difference between HB and SH is in their hyperparameters. HB has the same hyperparameters as SH (see the *Understanding SH* section) except for n_candidates. In HB, we don't have to choose the best value for n_candidates since it is calculated automatically within the HB algorithm.

Basically, HB works by running SH iteratively with variations of n_candidates and min_resources in each of the **brackets** (each SH run), starting from the combination of the highest possible value for n_candidates and the lowest possible value for min_resources, and going to the lowest possible value for n_candidates and the highest possible value for *resources* (see *Figure 6.8*). It's like a brute-force approach to try *almost* all of the possible combinations of n_candidates and min_resources.

SH Iteration	Bracket-1		Bracket-2		Bracket-3		Bracket-4	
	n_1	r_1	n_2	r_2	n_3	r_3	n_4	r_4
1	27	1	12	3	6	9	4	27
2	9	3	4	9	2	27		
3	3	9	1	27				
4	1	27						

Figure 6.8 – Illustration of the HB process. Here, nj and rj refer to n_candidates and min_resources for bracket-j, respectively

As illustrated in *Figure 6.8*, assume that we set `factor=3`, `min_resources=1`, `max_resources=27`, and `min_early_stopping=0`. As you can see, HB allocates the minimum amount of resources with the maximum number of candidates in the first bracket, while it allocates the maximum amount of resources with the minimum number of candidates in the last bracket. Again, each bracket refers to each SH run, meaning we are running SH four times in this illustration, where the last bracket is basically the same as performing random or grid search on a small hyperparameter space.

By testing *almost* all of the possible combinations of `n_candidates` and `min_resources`, HB is able to remove the trade-off in SH while also reducing the possibility of excluding better hyperparameters in the first iterations. However, this groundbreaking characteristic of HB *doesn't ensure that it will be always better than SH*. Why? Because HB hasn't actually tried all the possible combinations. We might find a better combination of `n_candidates` and `min_resources` values just by performing a single SH than all the possible combinations HB tried. However, this takes time and luck since we have to manually select the `n_candidates` and `min_resources` values.

> **Integration with Other Black-Box Methods**
>
> In the original paper on HB, the authors utilize random search for each SH run. However, as with SH, we can also integrate HB with other black-box methods.

The following procedure further states how HB works formally as a hyperparameter tuning method:

1. Split the full original dataset into train and test sets.

2. Define the hyperparameter space, `H`, with the accompanied distributions, and the objective function, `f`, based on the training set.

3. Define the `budget` resource. This is usually defined as the number of samples or training epochs.

4. Define the maximum resources, `max_resources`. This is usually defined as the total number of samples in the data or the total number of epochs.

5. Define the multiplier factor, `factor`, the minimum early stopping rate, `min_early_stopping`, and the minimum number of resources for all brackets, `min_resources`. Usually, `min_resources` is set to one, if the budget is defined as the number of samples.

6. Create a dictionary, `top_candidates`, that will be utilized to store the best-performing set of hyperparameters from each SH run.

7. Calculate the number of brackets, *nbrackets*, using the following formula:

$$n_{brackets} = 1 + floor\left(\log_{factor}\left(\frac{max\_resources}{min\_resources}\right)\right)$$

8. For each bracket-*j*, starting from j=1 until j=nbrackets, do the following:

    I.   Calculate the minimum number of resources to be used at the first iteration of SH for bracket-*j*, r_j , using the following formula:

    $$r_j = max\_resources \cdot factor^{-(n_{brackets}-j)}$$

    II. Calculate the initial number of hyperparameter candidates to be evaluated at the first iteration of SH for bracket-*j*, n_j, using the following formula:

    $$n_j = ceil\left(\frac{n_{brackets}}{n_{brackets} - j + 1} \cdot factor^{(n_{brackets}-j)}\right)$$

    III. Do *steps 7 – 11* from the SH procedure given in the *Understanding SH* section by utilizing r_j as the min_resources and n_j as the n_candidates for the current SH run, respectively. Other hyperparameters for SH, such as max_resources, min_early_ stopping, and factor, are inherited from HB.

    IV. Store the best set of hyperparameters output from the current SH run, along with the objective function score, in the top_candidates dictionary.

9. Select the best candidate that has the most optimal objective function score from the top_ candidates dictionary.

10. Train on the full training set using the best set of hyperparameters from *step 9*.

11. Evaluate the final trained model on the test set.

The following table summarizes the pros and cons of utilizing HB as a hyperparameter tuning method:

Pros	Cons
Same as SH (see *Figure 6.7*)	Higher computational cost
Able to remove the need for balancing the trade-off that SH has	

Figure 6.9 – Pros and cons of HB

It is worth noting that although HB can help us to deal with the trade-off of SH, it has a higher computational cost, since we have to run several SH rounds iteratively. It is even more costly when we are faced with a case where the bad and good hyperparameters cannot be easily distinguished with a small budget value. Why? The first several brackets of HB that utilize small budgets will result in a noisy estimation, since the relative rankings inside the SH iterations on smaller budgets do not reflect the actual relative rankings on higher budgets. In the most extreme case, the best set of hyperparameters will result from the last bracket (random search). If this is the case, then HB will run *nbrackets* times slower compared to random search.

In this section, we have discussed HB, what it is, how it works, and its pros and cons. We will discuss another interesting MFO method in the next section.

Understanding BOHB

Bayesian Optimization and Hyper Band (**BOHB**) is an extension of HB that is superior to CFS, SH, and HB, in terms of understanding the relationship between the hyperparameter candidates and the objective function. If CFS, SH, and HB are all part of the informed search group based on random search, BOHB is an informed search group that is based on the BO method. This means BOHB is able to decide which subspace needs to be searched based on previous experiences rather than luck.

As its name implies, BOHB is the combination of the BO and HB methods. While SH and HB can also be utilized with other black-box methods (see the *Understanding SH* and *Understanding HB* sections), BOHB is specifically designed to utilize a BO method in a way that can support HB. Furthermore, the BO method in BOHB also tracks all the previous evaluations on all budgets, so that it can serve as the base for future evaluations. Note that the BO method used in BOHB is the **multivariate TPE**, which is able to take into account the interdependencies among hyperparameters (see *Chapter 4, Exploring Bayesian Optimization*).

The main selling point of BOHB is its ability to achieve both a strong initial performance and a strong final performance. This can be easily seen in *Figure 6.10*, from the original BOHB paper (see the following note for details). BO (without performing metalearning) will outperform random search if we have more time to let it learn from previous experiences. If we don't have time, BO will deliver a similar or even worse performance compared to random search. On the other hand, HB performs much better than random search when we have limited time, but will perform similarly to random search if we allow more time for random search to explore the hyperparameter space. By combining the best of both worlds, BOHB is able to not only outperform random search in a limited time but also when given enough time for random search to catch up.

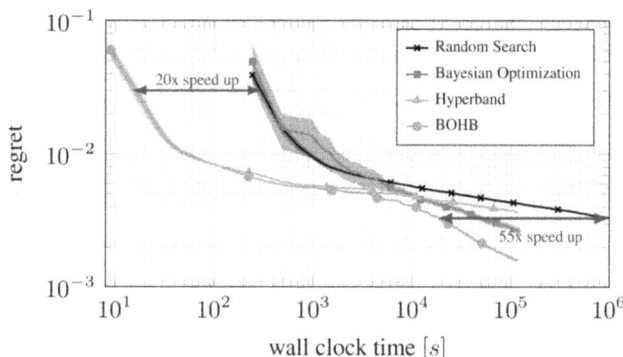

Figure 6.10 – Comparison between random search, BO, HB, and BOHB

> **The Original BOHB Paper**
>
> *BOHB: Robust and Efficient Hyperparameter Optimization at Scale* by Stefan Falkner, Aaron Klein, and Frank Hutter, Proceedings of the 35th International Conference on Machine Learning, PMLR 80:1437-1446, 2018 (http://proceedings.mlr.press/v80/falkner18a. html).

The following procedure further states how BOHB works formally as a hyperparameter tuning method. Note that BOHB and HB are very similar except that random search in HB is replaced by the combination of multivariate TPE and random search. Since HB just performs SH several times iteratively, the actual replacement is actually performed in each of the SH runs (each bracket) in HB.

Let's pick up from the previous instructions again.

6. *(The first six steps are the same as those in the Understanding HB section.)*

7. Define the probability of just performing a random search rather than fitting the multivariate TPE, *random_prob*.

8. Define the percentage of the good set of hyperparameters for the multivariate TPE fitting procedure, *top_n_percent*. (See *Chapter 4, Exploring Bayesian Optimization*.)

9. Define a dictionary, *candidates_dict*, that stores the budget/resources used in a particular SH iteration and the pairs of hyperparameter candidates and the objective function score as the key and value, respectively.

10. Define the minimum number of sets of hyperparameters that are randomly sampled before starting to fit the multivariate TPE, n_min. By default, we set n_min to match the number of hyperparameters in the space plus one.

11. For each bracket-*j*, starting from j=1 until j=nbrackets, do the following:

    I.   Calculate the minimum number of resources to be used on the first iteration of SH for bracket-*j*, r_j , using the following formula:

    $$r_j = max\_resources \cdot factor^{-(n_{brackets}-j)}$$

    II.   Calculate the initial number of hyperparameters candidates to be evaluated on the first iteration of SH for bracket-j, n_j , using the following formula:

    $$n_j = ceil\left(\frac{n_{brackets}}{n_{brackets}-j+1} \cdot factor^{(n_{brackets}-j)}\right)$$

    III. Perform steps 7 – 11 from the SH procedure stated in the Understanding SH section by utilizing r_j as min_resources and n_j as n_candidates for the current SH run, respectively, where step 9. I. is replaced with the following procedure:

    IV.   Generate a random number between zero and one from a uniform distribution, rnd.

V.   If `rnd<random_prod` or *models_dict* is empty, perform a random search to sample the initial hyperparameter candidates.

VI.   Count the number of sampled hyperparameters in `candidates_dict[`^{r}j`]`, and store it as `num_curr_candidates`.

VII.   If `num_curr_candidates < n_min`, then perform a random search to sample the initial hyperparameter candidates.

VIII.   Alternatively, utilize the multivariate TPE (see *Chapter 4, Exploring Bayesian Optimization*) to sample the initial hyperparameter candidates. Note that we always utilize multivariate TPE on the largest budget available in `candidates_dict`. The number of hyperparameter sets for both good and bad groups is defined based on the following formula:

$$n_{good} = \max\left(n\_min, floor\left(\frac{top\_n\_percent \cdot num\_curr\_candidates}{100}\right)\right)$$

$$n_{bad} = \max\left(n\_min, floor\left(\frac{(100 - top\_n\_percent) \cdot num\_curr\_candidates}{100}\right)\right)$$

IX.   Store the sampled initial hyperparameter candidates along with the objective function score (either from step ii, iv, or v) in `candidates_dict[`^{r}j`]`.

X.   Store the best set of hyperparameters output from the current SH run, along with the objective function score, in the `top_candidates` dictionary.

12. Select the best candidate that has the most optimal objective function score from the `top_candidates` dictionary.

13. Train on the full training set using the best set of hyperparameters from *step 14*.

14. Evaluate the final trained model on the test set.

Note that to ensure that BOHB tracks all of the evaluations on all budgets, we also need to store the hyperparameter candidates in each of the SH iterations for each HB bracket to `candidates_dict[budget]` along with their objective function score. Here, hyperparameter candidates in each of the SH iterations refer to *candidatesi*, while budget refers to *resourcesi* in *step 10* in the *Understanding SH* section, which also can be seen in the following figure:

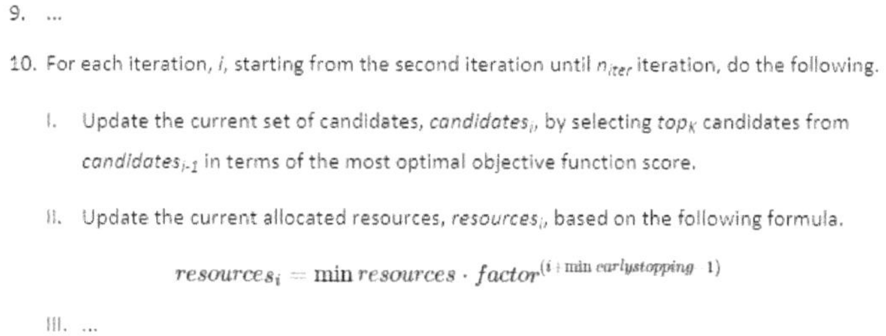

9.   ...

10. For each iteration, *i*, starting from the second iteration until *n*$_{iter}$ iteration, do the following.

   I.   Update the current set of candidates, *candidates*$_i$, by selecting *top*$_K$ candidates from *candidates*$_{i-1}$ in terms of the most optimal objective function score.

   II.   Update the current allocated resources, *resources*$_i$, based on the following formula.

$$resources_i = \min resources \cdot factor^{(i \,:\, \min\ earlystopping\ 1)}$$

   III.   ...

Figure 6.11 – BOHB tracks all the evaluations on all budgets

You may wonder whether it is possible for BOHB to take advantage of parallel resources since it utilizes a BO method that is notorious for not being able to exploit parallel computing resources. The answer is *yes, it is possible*! You can take advantage of parallel resources since in each of the BOHB iterations, specifically in the HB iterations, we can utilize more than one worker to evaluate multiple sets of hyperparameters, in parallel.

What about the sequential nature of the multivariate TPE utilized in BOHB? Yes, there may be some sequential processes that need to be performed inside the TPE model. However, BOHB limits the number of sets of hyperparameters given to the multivariate TPE so it might not take too much time. Furthermore, the limitation on the number of hyperparameter sets is actually specifically designed by the authors of BOHB. The following is a direct quote from the original paper on BOHB:

> *The parallelism in TPE is achieved by limiting the number of samples to*
> *optimize EI, purposefully not optimizing it fully to obtain diversity. This*
> *ensures that consecutive suggestions by the model are diverse enough to yield*
> *near-linear speedups when evaluated in parallel.*

It is also worth noting that we always utilize the multivariate TPE on the largest budget available to ensure that it is fitted on enough budget (high-fidelity) to minimize the chance of a noisy estimation. So, combined with the limitation on the number of hyperparameter sets passed to the TPE, we are trying to ensure that the multivariate TPE is fitted on the right number of hyperparameter sets.

The following table summarizes the pros and cons of utilizing BOHB as a hyperparameter tuning method:

Pros	Cons
Same as HB (see *Figure 6.9*)	Higher computational cost than HB
Able to decide which subspace needs to be searched based on previous experiences, not luck	
Achieves both a strong initial performance (inherited from HB) and a strong final performance (inherited from BO)	

Figure 6.12 – Pros and cons of BOHB

Just as HB may run *nbrackets* times slower compared to random search when we are faced with a situation where the bad and good hyperparameters cannot be easily distinguished with a small budget value, BOHB will also run *nbrackets* times slower compared to the vanilla BO, where we are faced with the same condition.

In this section, we have covered BOHB in detail, including what it is, how it works, and its pros and cons.

Summary

In this chapter, we have discussed the fourth of the four groups of hyperparameter tuning methods, called the MFO group. We have discussed MFO in general and what makes it different from black-box optimization methods, as well as discussing several variants, including CFS, SH, HB, and BOHB. We have seen the differences between them and the pros and cons of each. From now on, you should be able to explain MFO with confidence when someone asks you about it. You should also be able to debug and set up the most suitable configuration for the chosen method that suits your specific problem definition.

In the next chapter, we will begin implementing the various hyperparameter tuning methods that we have learned about so far using the scikit-learn package. We will become familiar with the scikit-learn package and learn how to utilize it in various hyperparameter tuning methods.

Section 2: The Implementation

In this section of the book, we will learn how to utilize several powerful packages to implement all of the discussed hyperparameter tuning methods in the previous section.

This section includes the following chapters:

7
Hyperparameter Tuning via Scikit

`scikit-learn` is one of the Python packages that is used the most by data scientists. This package provides a range of **Machine Learning** (**ML**)-related modules that are ready to be used with minimum effort, including for the task of hyperparameter tuning. One of the main selling points of this package is its consistent interface across many implemented classes, which almost every data scientist loves! Apart from `scikit-learn`, there are also other packages for the hyperparameter tuning task that are built on top of `scikit-learn` or mimic the interface of `scikit-learn`, such as `scikit-optimize` and `scikit-hyperband`.

In this chapter, we'll learn about all of the important things to do with `scikit-learn`, `scikit-optimize`, and `scikit-hyperband`, along with how to utilize them to implement the hyperparameter tuning methods that we learned about in the previous chapters. We'll start by walking through how to install each of the packages. Then, we'll learn not only how to utilize those packages with their default configurations but also discuss the available configurations along with their usage. Additionally, we'll discuss how the implementation of the hyperparameter tuning methods is related to the theory that we learned in previous chapters, as there might be some minor differences or adjustments made in the implementation.

Finally, equipped with the knowledge from previous chapters, you will also be able to understand what's happening if there are errors or unexpected results and understand how to set up the method configuration to match your specific problem.

In this chapter, we'll be covering the following main topics:

- Introducing scikit
- Implementing Grid Search
- Implementing Random Search
- Implementing Coarse-to-Fine Search

- Implementing Successive Halving

- Implementing Hyper Band

- Implementing Bayesian Optimization Gaussian Process

- Implementing Bayesian Optimization Random Forest

- Implementing Bayesian Optimization Gradient Boosted Trees

Technical requirements

We will learn how to implement various hyperparameter tuning methods with `scikit-learn`, `scikit-optimize`, and `scikit-hyperband`. To ensure that you can reproduce the code examples in this chapter, you will require the following:

- Python 3 (version 3.7 or above)

- An installed `Pandas` package (version 1.3.4 or above)

- An installed `NumPy` package (version 1.21.2 or above)

- An installed `Scipy` package (version 1.7.3 or above)

- An installed `Matplotlib` package (version 3.5.0 or above)

- An installed `scikit-learn` package (version 1.0.1 or above)

- An installed `scikit-optimize` package (version 0.9.0 or above)

- An installed `scikit-hyperband` package (directly cloned from the GitHub repository)

All of the code examples for this chapter can be found on GitHub at `https://github.com/PacktPublishing/Hyperparameter-Tuning-with-Python`.

Introducing Scikit

`scikit-learn`, which is commonly called **Sklearn**, is a very popular open source package in Python that is widely used for ML-related tasks, starting from data preprocessing, model training and evaluation, model selection, hyperparameter tuning, and more. One of the main selling points of the `sklearn` package is the consistency of its interface across many implemented classes.

For example, all of the implemented ML models, or **estimators**, in `sklearn` have the same `fit()` and `predict()` methods for fitting the model on the training data and evaluating the fitted model on the test data, respectively. When working with data preprocessors, or **transformers**, in `sklearn`, the typical method that every preprocessor has is the `fit()`, `transform()`, and `fit_transform()` methods for fitting the preprocessor, transforming new data with the fitted preprocessor, and fitting and directly transforming the data that is used to fit the preprocessor, respectively.

In *Chapter 1, Evaluating Machine Learning Models*, we learned how `sklearn` can be utilized to evaluate the performance of ML models through the concept of cross-validation, where the full data is split into several parts, such as train, validation, and test data. In *Chapters 3–6*, we always used the cross-validation score as our objective function. While we can manually perform hyperparameter tuning and calculate the cross-validation score based on the split data, `sklearn` provides dedicated classes for hyperparameter tuning that use the cross-validation score as the objective function during the tuning process. There are several hyperparameter tuning classes implemented in `sklearn`, such as `GridSearchCV`, `RandomizedSearchCV`, `HalvingGridSearchCV`, and `HalvingRandomSearchCV`.

Also, all of the hyperparameter tuning classes implemented in `sklearn` have a consistent interface. We can use the `fit()` method to perform hyperparameter tuning on the given data where the cross-validation score is used as the objective function. Then, we can use the `best_params_` attribute to get the best set of hyperparameters, the `best_score_` attribute to get the average cross-validated score from the best set of hyperparameters, and the `cv_results_` attribute to get the details of the hyperparameter tuning process, including but not limited to the objective function score for each tested set of hyperparameters in each of the folds.

To prevent data leakage when performing the data preprocessing steps (see *Chapter 1, Evaluating Machine Learning Models*), `sklearn` also provides a `Pipeline` object that can be used along with the hyperparameter tuning classes. This `Pipeline` object will ensure that any data preprocessing steps are only fitted based on the train set during the cross-validation. Essentially, this object is just *a chain of several* `sklearn` *transformers and estimators*, which has the same `fit()` and `predict()` method, just like a usual `sklearn` estimator.

While `sklearn` can be utilized for many ML-related tasks, `scikit-optimize`, which is commonly called **skopt**, is a package built on top of `sklearn` and can be utilized for implementing the **Sequential Model-Based Optimization** (**SMBO**) methods (see *Chapter 4, Exploring Bayesian Optimization*). `skopt` has a very similar interface to `sklearn`, so it will be very easy for you to get familiar with `skopt` once you are already familiar with `sklearn` itself. The main hyperparameter tuning class implemented in `skopt` is the `BayesSearchCV` class.

`skopt` provides four implementations for the optimizer within the `BayesSearchCV` class, namely **Gaussian Process** (**GP**), **Random Forest** (**RF**), **Gradient Boosted Regression Trees** (**GBRT**), and **Extra Trees** (**ET**). Furthermore, you can also use any other regressors from `sklearn` to be utilized as the optimizer. Note that, here, the optimizer refers to the **surrogate model** that we learned in *Chapter 4, Exploring Bayesian Optimization*. Additionally, `skopt` provides various implementations of the **acquisition function**, namely the **Expected Improvement** (**EI**), **Probability of Improvement** (**PI**), and **Lower Confidence Bound** (**LCB**) functions.

Last but not least, the `scikit-hyperband` package. Additionally, this package is built on top of `sklearn` and is specifically designed for the HB implementation. The hyperparameter tuning class implemented in this package is `HyperbandSearchCV`. It also has a very similar interface to `sklearn`.

As for `sklearn` and `skopt`, you can easily install them via `pip install`, just like you usually install other packages. As for `scikit-hyperband`, the author of the package didn't put this on **PyPI**, which means you have to install the package directly from the GitHub repository. Furthermore, at the time of writing, the last update of the GitHub repo (`https://github.com/thuijskens/scikit-hyperband`) was in 2020. There are several blocks of code that are no longer compatible with the newer version of `sklearn`. Luckily, there's a forked version (`https://github.com/louisowen6/scikit-hyperband`) of the original repo that works nicely with the newer version of `sklearn` (`1.0.1` or above). To install `scikit-hyperband`, please follow the following steps:

1. Clone `https://github.com/louisowen6/scikit-hyperband` to your loca machinel:

   ```
   git clone https://github.com/louisowen6/scikit-hyperband
   ```

2. Open the cloned repository:

   ```
   cd scikit-hyperband
   ```

3. Move the `hyperband` folder to your working directory:

   ```
   mv hyperband "path/to/your/working/directory"
   ```

Now that you are aware of the `scikit-learn`, `scikit-optimize`, and `scikit-hyperband` packages, in the following sections, we will learn how to utilize them to implement various hyperparameter tuning methods.

Implementing Grid Search

To implement **Grid Search** (see *Chapter 3, Exploring Exhaustive Search*), we can actually write our own code from scratch since it is just a simple nested `for loop` that tests all of the possible hyperparameter values in the search space. However, by using `sklearn`'s implementation of Grid Search, `GridSearchCV`, we can have a cleaner code since we just need to call a single line of code to instantiate the class.

Let's walk through an example of how we can utilize `GridSearchCV` to perform Grid Search. Note that, in this example, we are performing hyperparameter tuning on an RF model. We will utilize `sklearn`'s implementation of RF, `RandomForestClassifier`. The dataset used in this example is the *Banking Dataset – Marketing Targets* provided on Kaggle (`https://www.kaggle.com/datasets/prakharrathi25/banking-dataset-marketing-targets`).

> **Original Data Source**
> This data was first published in *A Data-Driven Approach to Predict the Success of Bank Telemarketing*, by Sérgio Moro, Paulo Cortez, and Paulo Rita, Decision Support Systems, Elsevier, 62:22–31, June 2014 (`https://doi.org/10.1016/j.dss.2014.03.001`).

This is a binary classification dataset with 16 features related to the marketing campaigns conducted by a bank institution. The target variable consists of two classes, *yes* or *no*, indicating whether the client of the bank has subscribed to a term deposit or not. Hence, the goal of training an ML model on this dataset is to identify whether a customer is potentially wanting to subscribe to the term deposit or not. For more details, you can refer to the description on the Kaggle page:

1. There are two datasets provided, namely the `train.csv` dataset and the `test.csv` dataset. However, we will not use the provided `test.csv` dataset since it is sampled directly from the train data. We will manually split `train.csv` into two subsets, namely the train set and the test set, using the help of the `train_test_split` function from `sklearn` (see *Chapter 1, Evaluating Machine Learning Models*). We will set the `test_size` parameter to `0.1`, meaning we will have `40,689` and `4,522` rows for the train set and the test set, respectively. The following code shows you how to load the data and perform the train set and the test set splitting:

```
import pandas as pd
from sklearn.model_selection import train_test_split
df = pd.read_csv("train.csv",sep=";")
df_train, df_test = train_test_split(df, test_size=0.1,
random_state=0)
```

Out of the 16 features provided in the data, there are 7 numerical features and 9 categorical features. As for the target class distribution, 12% of them are *yes* and 88% of them are *no*, for both train and test datasets. This means that we can't use accuracy as our metric since we have an imbalanced class problem—a situation where we have a very skewed distribution of the target classes. Instead, in this example, we will use the F1-score.

2. Before performing Grid Search, let's see how *RandomForestClassifier* with the default hyperparameter values work. Furthermore, let's also try to train our model on only those seven numerical features for now. The following code shows you how to get only numerical features, train the model on those features in the train set, and finally, evaluate the model on the test set:

```
import numpy as np
from sklearn.ensemble import RandomForestClassifier
from sklearn.metrics import f1_score
```

The `X_train_numerical` variable only stores numerical features from the train data:

```
X_train_numerical = df_train.select_dtypes(include=np.
number).drop(columns=['y'])
y_train = df_train['y']
```

The `X_test_numerical` variable only stores numerical features from the test data:

```
X_test_numerical = df_test.select_dtypes(include=np.
```

```
number).drop(columns=['y'])
y_test = df_test['y']
```

Fit the model on train data:

```
model = RandomForestClassifier(random_state=0)
model.fit(X_train_numerical,y_train)
```

Evaluate the model on the test data:

```
y_pred = model.predict(X_test_numerical)
print(f1_score(y_test, y_pred))
```

Based on the preceding code, we get around 0.436 for the F1-Score when testing our trained RF model on the test set. Remember that this is the result of only using numerical features and the default hyperparameters of the RandomForestClassifier.

3. Before performing Grid Search, we have to define the hyperparameter space in a dictionary of list format, where the keys refer to the name of the hyperparameters and the lists consist of all the values we want to test for each hyperparameter. Let's say we define the hyperparameter space for RandomForestClassifier as follows:

```
hyperparameter_space = {
"n_estimators": [25,50,100,150,200],
"criterion": ["gini", "entropy"],
"max_depth": [3, 5, 10, 15, 20, None],
"class_weight": ["balanced","balanced_subsample"],
"min_samples_split": [0.01,0.1,0.25,0.5,0.75,1.0],
}
```

4. Once we have defined the hyperparameter space, we can apply the GridSearchCV class to the train data, use the best set of hyperparameters to train a new model on the full train data, and then evaluate that final trained model on the test data, just as we learned in *Chapters 3–6*. The following code shows you how to do that:

```
from sklearn.model_selection import GridSearchCV
```

Initiate the model:

```
model = RandomForestClassifier(random_state=0)
```

Initiate the GridSearchCV class:

```
clf = GridSearchCV(model, hyperparameter_space,
                   scoring='f1', cv=5,
```

```
                              n_jobs=-1, refit = True)
```

Run the `GridSearchCV` class:

```
clf.fit(X_train_numerical, y_train)
```

Print the best set of hyperparameters:

```
print(clf.best_params_,clf.best_score_)
```

Evaluate the final trained model on the test data:

```
print(clf.score(X_test_numerical,y_test))
```

Look how clean our code is by utilizing `sklearn`'s implementation of Grid Search instead of writing our code from scratch! Notice that we just need to pass `sklearn`'s estimator and the hyperparameter space dictionary to the `GridSearchCV` class, and the rest will be handled by `sklearn`. In this example, we also pass several additional parameters to the class, such as `scoring='f1'`, `cv=5`, `n_jobs=-1`, and `refit=True`.

As its name suggests, the `scoring` parameter governs the scoring strategy that we want to use to evaluate our model during the cross-validation. While our objective function will always be the cross-validation score, this parameter controls what type of score we want to use as our metric. In this example, we are using the F1-score as our metric. However, you can also pass a custom callable function as the scoring strategy.

> **Available Scoring Strategies in Sklearn**
>
> You can refer to `https://scikit-learn.org/stable/modules/model_evaluation.html#scoring-parameter` for all of the implemented scoring strategies by `sklearn`, and refer to `https://scikit-learn.org/stable/modules/model_evaluation.html#scoring` if you want to implement your own custom scoring strategy.

The `cv` parameter indicates how many folds of cross-validation you want to perform. The `n_jobs` parameter controls how many jobs you want to run in parallel. If you decide to use all of the processors, you can simply set `n_jobs=-1`, just as we did in the example.

Last but not least, we have the `refit` parameter. This Boolean parameter is responsible for deciding whether at the end of the hyperparameter tuning process we want to refit our model on the full train set using the best set of hyperparameters or not. In this example, we set `refit=True`, meaning that `sklearn` will automatically refit our RF model on the full train set using the best set of hyperparameters. It is very important to retrain our model on the full train set after performing hyperparameter tuning since we only utilize subsets of the train set during the hyperparameter tuning process. There are several other parameters that you can control when initiating a `GridSearchCV` class. For more details, you can refer to the official page of `sklearn` (`https://scikit-learn.org/stable/modules/generated/sklearn.model_selection.GridSearchCV.html`).

Let's go back to our example. By performing Grid Search in the predefined hyperparameter space, we are able to get an F1-score of 0.495 when evaluated on the test set. The best set of hyperparameters is {'class_weight': 'balanced', 'criterion': 'entropy', 'min_samples_ split': 0.01, 'n_estimators': 150} with an objective function score of 0.493. Note that we can get the best set of hyperparameters along with its objective function score via the best_params_ and best_score_ attributes, respectively. Not bad! We get around 0.06 of improvement in the F1-score. However, note that we are still only using numerical features.

Next, we will try to utilize not only numerical features but also categorical features from our data. To be able to utilize those categorical features, we need to perform the **categorical encoding** preprocessing step. Why? Because ML models are not able to understand non-numerical features. Therefore, we need to convert those non-numerical features into numerical ones so that the ML model is able to utilize those features.

Remember that when we want to perform any data preprocessing steps, we have to be very careful with it to prevent any data leakage problem where we might introduce part of our test data into the train data (see *Chapter 1, Evaluating Machine Learning Models*). To prevent this problem, we can utilize the Pipeline object from sklearn. So, instead of passing an estimator to the GridSearchCV class, we can also pass a Pipeline object that consists of a chain of data preprocessors and an estimator:

1. Since, in this example, not all of our features are categorical and we only want to perform categorical encoding on those non-numerical features, we can utilize the ColumnTransformer class to specify which features we want to apply the categorical encoding step. Let's say we also want to perform a normalization preprocessing step on the numerical features. We can also pass those numerical features to the ColumnTransformer class along with the normalization transformer. Then, it will automatically apply the normalization step to only those numerical features. The following code shows you how to create such a Pipeline object with ColumnTransformer, where we use StandardScaler for the normalization step and OneHotEncoder for the categorical encoding step:

   ```
   from sklearn.preprocessing import StandardScaler,
   OneHotEncoder
   from sklearn.compose import ColumnTransformer
   from sklearn.pipeline import Pipeline
   ```

   Get list of numerical features and categorical features:

   ```
   numerical_feats = list(df_train.drop(columns='y').select_
   dtypes(include=np.number).columns)
   categorical_feats = list(df_train.drop(columns='y').
   select_dtypes(exclude=np.number).columns)
   ```

Initiate the preprocessor for numerical features and categorical features:

```
numeric_preprocessor = StandardScaler()
categorical_preprocessor = OneHotEncoder(handle_
unknown="ignore")
```

Delegate each preprocessor to the corresponding features:

```
preprocessor = ColumnTransformer(
    transformers=[
        ("num", numeric_preprocessor, numerical_feats),
        ("cat", categorical_preprocessor, categorical_
feats),
    ])
```

Create a pipeline of preprocessors and models. In this example, we named our pr-processing steps as *"preprocessor"* and the modeling step as *"model"*:

```
pipe = Pipeline(
    steps=[("preprocessor", preprocessor),
          ("model", RandomForestClassifier(random_
state=0))])
```

As you can see in the previous code blocks, the `ColumnTransformer` class is responsible for delegating each preprocessor to the corresponding features. Then, we can just reuse it for all of our preprocessing steps through a single preprocessor variable. Finally, we can create a pipeline consisting of the preprocessor variable and `RandomForestClassifier`. Note that within the `ColumnTransformer` class and the `Pipeline` class, we also have to provide the name of each preprocessor and step in the pipeline, respectively.

2. Now that we have defined the pipeline, we can see how our model performs on the test set (without hyperparameter tuning) by utilizing all of the features and preprocessors defined in the pipeline. The following code shows how we can directly use the pipeline to perform the same `fit()` and `predict()` methods as we did earlier:

```
pipe.fit(X_train_full,y_train)
y_pred = pipe.predict(X_test_full)
print(f1_score(y_test, y_pred))
```

Based on the preceding code, we get around 0.516 for the F1-score when testing our trained pipeline on the test set.

3. Next, we can start performing Grid Search over the pipeline, too. However, before we can do that, we need to redefine the hyperparameter space. We need to change the keys in the dictionary with the format of `<estimator_name_in_pipeline>__<hyperparameter_name>`. The following is the redefined version of our hyperparameter space:

```
hyperparameter_space = {
"model__n_estimators": [25,50,100,150,200],
"model__criterion": ["gini", "entropy"],
"model__class_weight": ["balanced", "balanced_
subsample"],
"model__min_samples_split": [0.01,0.1,0.25,0.5,0.75,1.0],
}
```

4. The following code shows you how to perform Grid Search over the pipeline instead of the estimator itself. Essentially, the code is the same as the previous version. The only difference is that we are performing the Grid Search over the pipeline and on *all* of the features in the data, not just the numerical features.

Initiate the `GridSearchCV` class:

```
clf = GridSearchCV(pipe, hyperparameter_space,
                   scoring = 'f1', cv=5,
                   n_jobs=-1, refit = True
```

Run the `GridSearchCV` class:

```
clf.fit(X_train_full, y_train)
```

Print the best set of hyperparameters:

```
print(clf.best_params_, clf.best_score_)
```

Evaluate the final trained model on the test data:

```
print(clf.score(X_test_full, y_test))
```

Based on the preceding code, we get around `0.549` for the F1-Score when testing our final trained RF model with the best set of hyperparameters on the test set. The best set of hyperparameters is {'model__class_weight': 'balanced_subsample', 'model__criterion': 'gini', 'model__min_samples_split': 0.01, 'model__n_estimators': 100} with an objective function score of 0.549.

It is worth noting that we can also *create a pipeline within a pipeline*. For example, we can create a pipeline for `numeric_preprocessor` that consists of a chain of missing value imputation and normalization modules. The following code shows how we can create such a pipeline. The `SimpleImputer` class is the missing value imputation transformer from `sklearn` that can help us to perform mean, median, mode, or constant imputation strategies if there are any missing values:

```
from sklearn.impute import SimpleImputer
numeric_preprocessor = Pipeline(
steps=[("missing_value_imputation",
SimpleImputer(strategy="mean")),       ("normalization",
StandardScaler())]
)
```

In this section, we have learned how to implement Grid Search in `sklearn` through the `GridSearchCV` class, starting from defining the hyperparameter space, setting each important parameter of the `GridSearchCV` class, learning how to utilize the `Pipeline` and `ColumnTransformer` classes to prevent data leakage issues, and learning how to create a pipeline within the pipeline.

In the next section, we will learn how to implement Random Search in `sklearn` via `RandomizedSearchCV`.

Implementing Random Search

Implementing **Random Search** (see *Chapter 3, Exploring Exhaustive Search*) in `sklearn` is very similar to implementing Grid Search. The main difference is that we have to provide the number of trials or iterations since Random Search will not try all of the possible combinations in the hyperparameter space. Additionally, we have to provide the accompanying distribution for each of the hyperparameters when defining the search space. In `sklearn`, Random Search is implemented in the `RandomizedSearchCV` class.

To understand how we can implement Random Search in `sklearn`, let's use the same example from the *Implementing Grid Search* section. Let's directly try using all of the features available in the dataset. All of the pipeline creation processes are exactly the same, so we will directly jump into the process of how to define the hyperparameter space and the `RandomizedSearchCV` class. The following code shows you how to define the accompanying distribution for each of the hyperparameters in the space:

```
from scipy.stats import randint, truncnorm
hyperparameter_space = {
"model__n_estimators": randint(5, 200),
"model__criterion": ["gini", "entropy"],
"model__class_weight": ["balanced","balanced_subsample"],
"model__min_samples_split": truncnorm(a=0,b=0.5,loc=0.005,
scale=0.01),
}
```

As you can see, the hyperparameter space is quite different from the one that we defined previously in the *Implementing Grid Search* section. Here, we are also specifying the distribution for each of the hyperparameters, where `randint` and `truncnorm` are utilized for the `n_estimators` and `min_samples_split` hyperparameters. As for `criterion` and `class_weight`, we are still using the same configuration as the previous search space. Note that *by not specifying any distribution* means we are *applying uniform distribution* to the hyperparameter, where all of the values will have the same probability to be tested.

Essentially, the `randint` distribution is just a uniform distribution for discrete variables, while `truncnorm` stands for truncated normal distribution, which, as its name suggests, is a modified normal distribution bounded on a particular range. In this example, the range is bounded on a range from `a=0` and `b=0.5`, with a mean of `loc=0.005` and a standard deviation of `scale=0.01`.

> **Distribution for Hyperparameters**
>
> There are many other available distributions that you can utilize. `sklearn` accepts all distributions that have the `rvs` method, as in the distribution implementation from `Scipy`. Essentially, this method is just a method to sample a value from the specified distribution. For more details, please refer to the official documentation page of `Scipy` (`https://docs.scipy.org/doc/scipy/reference/stats.html#probability-distributions`).

When initiating the `RandomizedSearchCV` class, we also have to define the `n_iter` and `random_state` parameters, which refer to the number of iterations and the random seed, respectively. The following code shows you how to perform Random Search over the same pipeline defined in the *Implementing Grid Search* section. In contrast with the example in the *Implementing Grid Search* section, which only performs `120` iterations of Grid Search, here, we perform `200` iterations of random search since we set `n_iter=200`. Additionally, we have a bigger hyperparameter space since we increase the granularity of the `n_estimators` and `min_samples_split` hyperparameter values:

```
from sklearn.model_selection import RandomizedSearchCV
```

Initiate the `RandomizedSearchCV` class:

```
clf = RandomizedSearchCV(pipe, hyperparameter_space,
                        n_iter = 200, random_state = 0,
                        scoring = 'f1', cv=5,
                        n_jobs=-1, refit = True)
```

Run the `RandomizedSearchCV` class:

```
clf.fit(X_train_full, y_train)
```

Print the best set of hyperparameters:

```
print(clf.best_params_, clf.best_score_)
```

Evaluate the final trained model on the test data:

```
print(clf.score(X_test_full, y_test))
```

Based on the preceding code, we get around `0.563` for the F1-score when testing our final trained RF model with the best set of hyperparameters on the test set. The best set of hyperparameters is `{'model__class_weight': 'balanced_subsample', 'model__criterion': 'entropy', 'model__min_samples_split': 0.005155815445940717, 'model__n_estimators': 187}` with an objective function score of `0.562`.

In this section, we have learned how to implement Random Search in `sklearn` through the `RandomizedSearchCV` class, starting from defining the hyperparameter space to setting each important parameter of the `RandomizedSearchCV` class. In the next section, we will learn how to perform CFS with `sklearn`.

Implementing Coarse-to-Fine Search

Coarse-to-Fine Search (CFS) is part of the Multi-Fidelity Optimization group that utilizes Grid Search and/or Random Search during the hyperparameter tuning process (see *Chapter 6, Exploring Multi-Fidelity Optimization*). Although CFS is not implemented directly in the `sklearn` package, you can find the implemented custom class, `CoarseToFineSearchCV`, in the repo mentioned in the *Technical Requirements* section.

Let's use the same example and hyperparameter space as in the *Implementing Random Search* section, to see how `CoarseToFineSearchCV` works in practice. Note that this implementation of CFS only utilizes Random Search and uses the top *N* percentiles scheme to define the promising subspace in each iteration, similar to the example shown in *Chapter 6*. However, *you can edit the code based on your own preference* since CFS is a very simple method with customizable modules.

The following code shows you how to perform CFS with the `CoarseToFineSearchCV` class. It is worth noting that this class has very similar parameters to the `RandomizedSearchCV` class, with several additional parameters. The `random_iters` parameter controls the number of iterations for each random search trial, `top_n_percentile` controls the *N* value within the top N percentiles promising subspace definition (see *Chapter 6*), `n_iter` defines the number of CFS iterations to be performed, and `continuous_hyperparams` stores the list of continuous hyperparameters in the predefined space.

Initiate the `CoarseToFineSearchCV` class:

```
clf = CoarseToFineSearchCV(pipe, hyperparameter_space,
random_iters=25, top_n_percentile=50, n_iter=10,
continuous_hyperparams=['model__min_samples_split'],
random_state=0, scoring='f1', cv=5,
n_jobs=-1, refit=True)
```

Run the `CoarseToFineSearchCV` class:

```
clf.fit(X_train_full, y_train)
```

Print the best set of hyperparameters:

```
print(clf.best_params_, clf.best_score_)
```

Evaluate the final trained model on the test data:

```
y_pred = clf.predict(X_test_full)
print(f1_score(y_test, y_pred))
```

Based on the preceding code, we get around 0.561 for the F1-score when testing our final trained RF model with the best set of hyperparameters on the test set. The best set of hyperparameters is `{'model__class_weight': 'balanced_subsample', 'model__criterion': 'entropy', 'model__min_samples_split': 0.005867409821769845, 'model__n_estimators': 106}` with an objective function score of 0.560.

In this section, we have learned how to implement CFS using a custom class on top of `sklearn` through the `CoarseToFineSearchCV` class. In the next section, we will learn how to perform SH with `sklearn`.

Implementing Successive Halving

Similar to CFS, **Successive Halving (SH)** is also part of the Multi-Fidelity Optimization group (see *Chapter 6*). There are two implementations of SH in `sklearn`, namely `HalvingGridSearchCV` and `HalvingRandomSearchCV`. As their names suggest, the former class is an implementation of SH that utilizes Grid Search in each of the SH iterations, while the latter utilizes Random Search.

By default, SH implementations in `sklearn` use the number of samples, or *n_samples*, as the definition of the budget or resource in SH. However, it is also possible to define a budget with other definitions. For example, we can use `n_estimators` in RF as the budget, instead of using the number of samples. It is worth noting that we cannot use `n_estimators`, or any other hyperparameters, to define the budget if it is part of the hyperparameter space.

Both `HalvingGridSearchCV` and `HalvingRandomSearchCV` have similar standard SH parameters to control how the SH iterations will work, such as the `factor` parameter, which refers to the multiplier factor for SH, `resource`, which refers to what definition of budget we want to use, `max_resources` refers to the maximum budget or resource, and `min_resources`, which refers to the minimum number of resources to be used at the first iteration. By default, the `max_resources` parameter is set to *auto*, meaning it will use the total number of samples that we have when `resource='n_samples'`. On the other hand, `sklearn` implemented a heuristic to define the default value for the `min_resources` parameter, referred to as *smallest*. This heuristic will ensure that we have a small value of `min_resources`.

Specific for `HalvingRandomSearchCV`, there is also the `n_candidates` parameter that refers to the initial number of candidates to be evaluated at the first iteration. Note that this parameter is not available in `HalvingGridSearchCV` since it will automatically evaluate all of the hyperparameter candidates in the predefined space. It is worth noting that `sklearn` implemented a strategy, called *exhaust*, to define the default value of the `n_candidates` parameter. This strategy ensures that we evaluate enough candidates at the first iteration so that we can utilize as many resources as possible at the last SH iteration.

Besides those standard SH parameters, both of the classes also have the `aggressive_elimination` parameter, which can be utilized when we have a low number of resources. If this Boolean parameter is set to `True`, `sklearn` will automatically rerun the first SH iteration several times until the number of candidates is small enough. The goal of this parameter is to ensure that we only evaluate a maximum of `factor` candidates in the last SH iteration. Note that this parameter is only implemented in `sklearn`, the original SH doesn't introduce this strategy as part of the tuning method (see *Chapter 6*).

Similar to `GridSearchCV` and `RandomizedSearchCV`, `HalvingGridSearchCV` and `HalvingRandomSearchCV` also have the usual default `sklearn` parameters for hyperparameter tuning, such as `cv`, `scoring`, `refit`, `random_state`, and `n_jobs`.

> **Experimental Features of SH in sklearn**
>
> It is worth noting that as per `version 1.0.2` of `sklearn`, the SH implementations are still in the experimental phase. This means that there might be changes in the implementation or interface of the classes without any depreciation cycle.

The following code shows how `HalvingRandomSearchCV` works with its default SH parameters. Note that we still use the same example and hyperparameter space as in the *Implementing Random Search* section. It is also worth noting that we only use the `HalvingRandomSearchCV` class in this example since `HalvingGridSearchCV` has a very similar interface:

```
from sklearn.experimental import enable_halving_search_cv
from sklearn.model_selection import HalvingRandomSearchCV
```

Initiate the `HalvingRandomSearchCV` class:

```
clf = HalvingRandomSearchCV(pipe, hyperparameter_space,
                            factor=3,
 aggressive_elimination=False,
                            random_state = 0,
                            scoring = 'f1', cv=5,
                            n_jobs=-1, refit = True)
```

Run the `HalvingRandomSearchCV` class:

```
clf.fit(X_train_full, y_train)
```

Print the best set of hyperparameters:

```
print(clf.best_params_, clf.best_score_)
```

Evaluate the final trained model on the test data:

```
print(clf.score(X_test_full, y_test))
```

Based on the preceding code, we get around 0.556 for the F1-score when testing our final trained RF model with the best set of hyperparameters on the test set. The best set of hyperparameters is `{'model__class_weight': 'balanced_subsample', 'model__criterion': 'entropy', 'model__min_samples_split': 0.007286406330027324, 'model__n_estimators': 42}` with an objective function score of 0.565.

1. The following code shows you how to generate a figure that shows the tuning process in each SH iteration:

    ```
    import matplotlib.pyplot as plt
    ```

    Get the fitting history of each trial:

    ```
    results = pd.DataFrame(clf.cv_results_)
    results["params_str"] = results.params.apply(str)
    results.drop_duplicates(subset=("params_str", "iter"),
    inplace=True)
    mean_scores = results.pivot(
    index="iter", columns="params_str", values="mean_test_
    score")
    ```

Plot the fitting history for each trial:

```
fig, ax = plt.subplots(figsize=(16,16))
ax = mean_scores.plot(legend=False, alpha=0.6, ax=ax)
labels = [
    f"Iteration {i+1}\nn_samples={clf.n_resources_[i]}\
nn_candidates={clf.n_candidates_[i]}"
    for i in range(clf.n_iterations_)]
ax.set_xticks(range(clf.n_iterations_))
ax.set_xticklabels(labels, rotation=0,
multialignment="left",size=16)
ax.set_title("F1-Score of Candidates over
Iterations",size=20)
ax.set_ylabel("5-Folds Cross Validation F1-Score",
fontsize=18)
ax.set_xlabel("")
plt.tight_layout()
plt.show()
```

2. Based on the preceding code, we get the following figure:

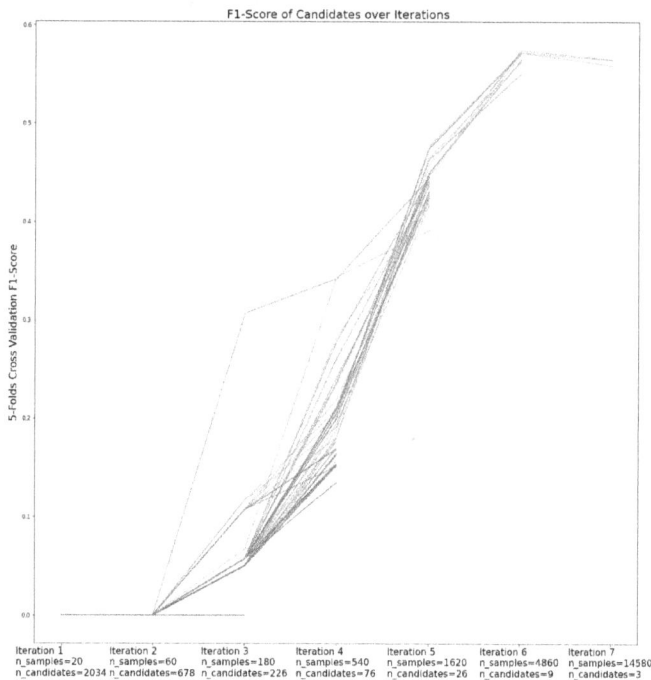

Figure 7.1 – The SH hyperparameter tuning process

Based on *Figure 7.1*, we can see that we only utilized around 14,000 samples in the last iteration while we have around 40,000 samples in our training data. Indeed, this is not an ideal case since there are too many samples not being utilized in the last SH iteration. We can change the default value of the SH parameters set by `sklearn` to ensure that we utilize as many resources as possible at the last iteration, through the `min_resources` and `n_candidates` parameters.

In this section, we have learned how to implement SH in `sklearn` through the `HalvingRandomSearchCV` and `HalvingGridSearchCV` classes. We have also learned all of the important parameters available for both classes. In the next section, we will learn how to perform HB with `scikit-hyperband`.

Implementing Hyper Band

The extension of Successive Halving, the **Hyper Band (HB)** method (see *Chapter 6*), is implemented in the `scikit-hyperband` package. This package is built on top of `sklearn`, which means it also provides a very similar interface for `GridSearchCV`, `RandomizedSearchCV`, `HalvingGridSearchCV`, and `HalvingRandomSearchCV`.

In contrast with the default SH budget definition in the `sklearn` implementation, *Scikit-Hyperband defines the budget* as the number of estimators, *n_estimators*, in an ensemble of trees, or the number of iterations for estimators trained with stochastic gradient descent, such as the XGBoost algorithm. Additionally, we can use any other hyperparameters that exist in the estimator as the budget definition. However, `scikit-hyperband` *doesn't allow us to use the number of samples as the budget definition*.

Let's use the same example as in the *Implementing Successive Halving* section, but with a different hyperparameter space. Here, we use the number of estimators, *n_estimators*, as the resource, which means we have to take out this hyperparameter from our search space. Note that you also have to remove any other hyperparameters from the space when you use it as the resource definition, just like in the `sklearn` implementation of SH.

The following code shows you how `HyperbandSearchCV` works. The `resource_param` parameter refers to the hyperparameter that you want to use as the budget definition. The `eta` parameter is actually the same as the factor parameter in the `HalvingRandomSearchCV` or `HalvingGridSearchCV` classes, which refers to the multiplier factor for each SH run. The `min_iter` and `max_iter` parameters refer to the minimum and maximum resources for all brackets. Note that there's no automatic strategy like in the `sklearn` implementation of SH for setting the value of the `min_iter` and `max_iter` parameters.

The remaining `HyperbandSearchCV` parameters are similar to any other `sklearn` implementation of the hyperparameter tuning methods. It is worth noting that the HB implementation used in this book is the modified version of the `scikit-hyperband` package. Please check the following folder in the book's GitHub repo (`https://github.com/PacktPublishing/Hyperparameter-Tuning-with-Python/tree/main/hyperband`):

```
from hyperband import HyperbandSearchCV
```

Initiate the `HyperbandSearchCV` class:

```
clf = HyperbandSearchCV(pipe, hyperparameter_space,
                        resource_param='model__n_estimators',
                        eta=3, min_iter=1, max_iter=100,
                        random_state = 0,
                        scoring = 'f1', cv=5,
                        n_jobs=-1, refit = True)
```

Run the `HyperbandSearchCV` class:

```
clf.fit(X_train_full, y_train)
```

Print the best set of hyperparameters:

```
print(clf.best_params_, clf.best_score_)
```

Evaluate the final trained model on the test data:

```
print(clf.score(X_test_full, y_test))
```

Based on the preceding code, we get around `0.569` in F1-score when testing our final trained RF model with the best set of hyperparameters on the test set. The best set of hyperparameters is `{'model__class_weight': 'balanced', 'model__criterion': 'entropy', 'model__min_samples_split': 0.0055643644642829684, 'model__n_estimators': 33}` with an objective function score of `0.560`. Note that although we remove `model__n_estimators` from the search space, `HyperbandSearchCV` still outputs the best value for this hyperparameter by choosing from the best bracket.

In this section, we have learned how to implement HB using the help of the `scikit-hyperband` package along with all of the important parameters available for the `HyperbandSearchCV` class. In the next section, we will learn how to perform Bayesian Optimization with `scikit-optimize`.

Implementing Bayesian Optimization Gaussian Process

Bayesian Optimization Gaussian Process (BOGP) is one of the variants of the Bayesian Optimization hyperparameter tuning group (see *Chapter 4, Exploring Bayesian Optimization*). To implement BOGP, we can utilize the `skopt` package. Similar to `scikit-hyperband`, this package is also built on top of the `sklearn` package, which means the interface for the implemented Bayesian Optimization tuning class, `BayesSearchCV`, is very similar to `GridSearchCV`, `RandomizedSearchCV`, `HalvingGridSearchCV`, `HalvingRandomSearchCV`, and `HyperbandSearchCV`.

However, unlike `sklearn` or `scikit-hyperband`, which works well directly with the distribution implemented in `scipy`, in `skopt`, we can only use the wrapper provided by the package when defining the hyperparameter space. The wrappers are defined within the `skopt.space.Dimension` instances and consist of three types of dimensions, such as `Real`, `Integer`, and `Categorical`. Within each of these dimension wrappers, `skopt` actually uses the same distribution from the `scipy` package.

By default, the `Real` dimension only supports the `uniform` and `log-uniform` distributions and can take any real/numerical value as the input. As for the `Categorical` dimension, this wrapper can only take categorical values as the input, as implied by its name. It will automatically convert categorical values into integers or even real values, which means we can also utilize categorical hyperparameters for BOGP! Although we can do this, remember that BOGP only works best for the actual real variables (see *Chapter 4, Exploring Bayesian Optimization*). Finally, we have the `Integer` dimension wrapper. By default, this wrapper only supports `uniform` and `log-uniform` distributions for integer formatting. The `uniform` distribution will utilize the `randint` distribution from `scipy`, while the `log-uniform` distribution is exactly the same as the one that is used in the Real wrapper.

It is worth noting that we can write our own wrapper for other distributions too; for example, the `truncnorm` distribution that we use in all of our earlier examples. In fact, you can find the custom `Real` wrapper that consists of the `truncnorm`, `uniform`, and `log-uniform` distributions in the repo mentioned in the *Technical Requirements* section. The following code shows you how we can define the hyperparameter space for `BayesSearchCV`. Note that we are still using the same example and hyperparameter space as the *Implementing Random Search* section. Here, `Integer` and `Categorical` are the original wrappers provided by `skopt`, while the `Real` wrapper is the custom wrapper that consists of the `truncnorm` distribution, too:

```
from skopt.space import *
hyperparameter_space = {
"model__n_estimators": Integer(low=5, high=200),
"model__criterion": Categorical(["gini", "entropy"]),
"model__class_weight": Categorical(["balanced","balanced_
subsample"]),
"model__min_samples_split":
```

```
Real(low=0,high=0.5,prior="truncnorm",
                        **{"loc":0.005,"scale":0.01})
}
```

All of the parameters of the `BayesSearchCV` class are very similar to the `GridSearchCV`, `RandomizedSearchCV`, `HalvingGridSearchCV`, `HalvingRandomSearchCV`, or `HyperbandSearchCV`. The only specific parameters for `BayesSearchCV` are the `n_iter` and `optimizer_kwargs` which refer to the total number of trials to be performed and the parameter that consists of all related parameters for the `Optimizer`, respectively. Here, the `Optimizer` is a class that represents each of the Bayesian Optimization steps, starting from initializing the initial points, fitting the surrogate model, sampling the next set of hyperparameters using the help of the acquisition function, and optimizing the acquisition function (see *Chapter 4*).

There are several parameters available that we can pass to the `optimizer_kwargs` dictionary. The `base_estimator` parameter refers to the type of surrogate model to be used. `skopt` has prepared several surrogate models with default setups, including the Gaussian Process or *GP*. The `n_initial_points` parameter refers to the number of random initial points before the actual Bayesian Optimization steps begin. The `initial_point_generator` parameter refers to the initialization method to be used. By default, `skopt` will initialize them randomly. However, you can also change the initialization method to *lhs*, *sobol*, *halton*, *hammersly*, or *grid*.

As for the type of acquisition function to be used, by default, `skopt` will use *gp_hedge*, which is an acquisition function that will automatically choose either one of the **Lower Confidence Bound (LCB)**, **Expected Improvement (EI)**, or **Probability of Improvement (PI)** based on the probability. However, we can also choose to use each of those acquisition functions independently, by setting the `acq_func` parameter to *LCB*, *EI*, and *PI*, respectively. As explained in *Chapter 4*, besides choosing what acquisition function needs to be used, we also have to define what kind of optimizer to be utilized for the acquisition function itself. There are two options for the acquisition function's optimizer provided by `skopt`, namely random sampling (*sampling*) and *lbfgs*, or the type of second-order optimization strategy mentioned in *Chapter 4*. By default, `skopt` sets the `acq_optimizer` parameter to *auto*, which will choose automatically when to use the *sampling* or *lbfgs* optimization methods.

Finally, we can also pass the `acq_func_kwargs` parameter within the `optimizer_kwargs` parameter. We can pass all parameters related to the acquisition function to this `acq_func_kwargs` parameter; for example, the `xi` parameter that controls the exploration and exploitation behavior of the BOGP, as explained in *Chapter 4*. While the `xi` parameter is responsible for controlling the exploration versus exploitation trade-off for EI and PI acquisition functions, there is also another parameter called `kappa`, which is responsible for the same task as the LCB acquisition function. The higher the value of `xi` or `kappa` means that we are favoring exploration over exploitation, and vice versa. For more information about all of the parameters that are available in the `BayesSearchCV` class, you can refer to the official API reference of the `skopt` package (`https://scikit-optimize.github.io/stable/modules/classes.html`).

The following code shows how we can utilize `BayesSearchCV` to perform BOGP on the same example as the *Implementing Random Search* section:

```
from skopt import BayesSearchCV
```

Initiate the BayesSearchCV class:

```
clf = BayesSearchCV(pipe, hyperparameter_space, n_iter=50,
optimizer_kwargs={"base_estimator":"GP",
                  "n_initial_points":10,
                  "initial_point_generator":"random",
                  "acq_func":"EI",
                  "acq_optimizer":"auto",
                  "n_jobs":-1,
                  "random_state":0,
                  "acq_func_kwargs": {"xi":0.01}
                  },
random_state = 0,
scoring = 'f1', cv=5,
n_jobs=-1, refit = True)
```

Run the BayesSearchCV class:

```
clf.fit(X_train_full, y_train)
```

Print the best set of hyperparameters:

```
print(clf.best_params_, clf.best_score_)
```

Evaluate the final trained model on the test data:

```
print(clf.score(X_test_full, y_test))
```

Based on the preceding code, we get around 0.539 for the F1-Score when testing our final trained RF model with the best set of hyperparameters on the test set. The best set of hyperparameters is {'model__class_weight': 'balanced', 'model__criterion': 'entropy', 'model__min_samples_split': 0.02363008892366518, 'model__n_estimators': 94} with an objective function score of 0.530.

In this section, we have learned how to implement BOGP in `skopt` along with all of the important parameters available for the `BayesSearchCV` class. It is worth noting that `skopt` also has experiment tracking modules that include several native supports for plotting the result. We will learn more about those modules in *Chapter 13, Tracking Hyperparameter Tuning Experiments*. In the next section, we will learn how to perform another variant of Bayesian Optimization that utilizes RF as its surrogate model with `skopt`.

Implementing Bayesian Optimization Random Forest

Bayesian Optimization Random Forest (BORF) is another variant of Bayesian Optimization hyperparameter tuning methods that utilize RF as the surrogate model. Note that this variant is different from **Sequential Model Algorithm Configuration (SMAC)** although both of them utilize RF as the surrogate model (see *Chapter 4, Exploring Bayesian Optimization*).

Implementing BORF with `skopt` is actually very similar to implementing BOGP as discussed in the previous section. We just need to change the `base_estimator` parameter within `optimizer_kwargs` to *RF*. Let's use the same example as in the *Implementing Bayesian Optimization Gaussian Process* section, but change the acquisition function from *EI* to *LCB*. Additionally, let's change the `xi` parameter in the `acq_func_kwargs` to *kappa* since we are using *LCB* as our acquisition function. Note that we can also still use the same acquisition function. The changes made here just to show how you can interact with the interface of the `BayesSearchCV` class:

```
from skopt import BayesSearchCV
```

Initiate the `BayesSearchCV` class:

```
clf = BayesSearchCV(pipe, hyperparameter_space, n_iter=50,
optimizer_kwargs={"base_estimator":"RF",
                  "n_initial_points":10,
                  "initial_point_generator":"random",
                  "acq_func":"LCB",
                  "acq_optimizer":"auto",
                  "n_jobs":-1,
                  "random_state":0,
                  "acq_func_kwargs": {"kappa":1.96}
                  },
random_state = 0,
scoring = 'f1', cv=5,
n_jobs=-1, refit = True)
```

Run the `BayesSearchCV` class:

```
clf.fit(X_train_full, y_train)
```

Print the best set of hyperparameters:

```
print(clf.best_params_, clf.best_score_)
```

Evaluate the final trained model on the test data.

```
print(clf.score(X_test_full, y_test))
```

Based on the preceding code, we get around 0.617 for the F1-score when testing our final trained RF model with the best set of hyperparameters on the test set. The best set of hyperparameters is {`'model__class_weight'`: `'balanced_subsample'`, `'model__criterion'`: `'gini'`, `'model__min_samples_split'`: 0.00043534042560206855, `'model__n_estimators'`: 85} with an objective function score of 0.616.

In this section, we have learned how to implement BORF in `skopt` through the `BayesSearchCV` class. In the next section, we will learn how to perform another variant of Bayesian Optimization, which utilizes Gradient Boosted Trees as its surrogate model with `skopt`.

Implementing Bayesian Optimization Gradient Boosted Trees

Bayesian Optimization Gradient Boosted Trees (**BOGBRT**) is another variant of Bayesian Optimization that utilizes Gradient Boosted Trees as a surrogate model. Note that there will be endless variants of Bayesian Optimization that we can implement in `skopt` since we can just pass any other regressors from `sklearn` to be utilized as the `base_estimator` parameter. However, *GBRT* is part of the default surrogate model with predefined default hyperparameter values from the `skopt` package.

Similar to the *Implementing Bayesian Optimization Random Forest* section, we can just change the `base_estimator` parameter within `optimizer_kwargs` to *GBRT*. The following code shows you how to implement BOGBRT in `skopt`:

```
from skopt import BayesSearchCV
```

Initiate the `BayesSearchCV` class:

```
clf = BayesSearchCV(pipe, hyperparameter_space, n_iter=50,
optimizer_kwargs={"base_estimator":"GBRT",
                  "n_initial_points":10,
                  "initial_point_generator":"random",
                  "acq_func":"LCB",
                  "acq_optimizer":"auto",
                  "n_jobs":-1,
                  "random_state":0,
                  "acq_func_kwargs": {"kappa":1.96}
                  },
random_state = 0,
scoring = 'f1', cv=5,
n_jobs=-1, refit = True)
```

Run the `BayesSearchCV` class:

```
clf.fit(X_train_full, y_train)
```

Print the best set of hyperparameters:

```
print(clf.best_params_, clf.best_score_)
```

Evaluate the final trained model on the test data:

```
print(clf.score(X_test_full, y_test))
```

Based on the preceding code, we get around 0.611 for the F1-Score when testing our final trained RF model with the best set of hyperparameters on the test set. The best set of hyperparameters is { 'model__class_weight': 'balanced_subsample', 'model__criterion': 'gini', 'model__min_samples_split': 0.0005745541104096049, 'model__n_estimators': 143} with an objective function score of 0.618.

In this section, we have learned how to implement BOGBRT in skopt through the `BayesSearchCV` class by using the same example as in the *Implementing Bayesian Optimization Random Forest* section.

Summary

In this chapter, we have learned all the important things about the `scikit-learn`, `scikit-optimize`, and `scikit-hyperband` packages for hyperparameter tuning purposes. Additionally, we have learned how to implement various hyperparameter tuning methods using the help of those packages, along with understanding each of the important parameters of the classes and how are they related to the theory that we have learned in the previous chapters. From now on, you should be able to utilize these packages to implement your chosen hyperparameter tuning method and, ultimately, boost the performance of your ML model. Equipped with the knowledge from *Chapters 3–6*, you will also be able to understand what's happening if there are errors or unexpected results and how to set up the method configuration to match your specific problem.

In the next chapter, we will learn about the Hyperopt package and how to utilize it to perform various hyperparameter tuning methods. The goal of the next chapter is similar to this chapter, that is, to be able to utilize the package for hyperparameter tuning purposes and understand each of the parameters of the implemented classes.

8

Hyperparameter Tuning via Hyperopt

Hyperopt is an optimization package in Python that provides several implementations of hyperparameter tuning methods, including **Random Search**, **Simulated Annealing (SA)**, **Tree-Structured Parzen Estimators (TPE)**, and **Adaptive TPE (ATPE)**. It also supports various types of hyperparameters with ranging types of sampling distributions.

In this chapter, we'll introduce the Hyperopt package, starting with its capabilities and limitations, how to utilize it to perform hyperparameter tuning, and all the other important things you need to know about Hyperopt. We'll learn not only how to utilize Hyperopt to perform hyperparameter tuning with its default configurations but also discuss the available configurations, along with their usage. Moreover, we'll discuss how the implementation of the hyperparameter tuning methods is related to the theory that we learned about in the previous chapters, since there some minor differences or adjustments may have been made in the implementation.

By the end of this chapter, you will be able to understand all the important things you need to know about Hyperopt and be able to implement various hyperparameter tuning methods available in this package. You'll also be able to understand each of the important parameters of their classes and how they are related to the theory that we learned about in the previous chapters. Finally, equipped with the knowledge from previous chapters, you will be able to understand what's happening if there are errors or unexpected results, as well as how to set up the method configuration so that it matches your specific problem.

The following topics will be covered in this chapter:

- Introducing Hyperopt
- Implementing Random Search
- Implementing Tree-Structured Parzen Estimators
- Implementing Adaptive Tree-Structured Parzen Estimators
- Implementing simulated annealing

Technical requirements

In this chapter, we will learn how to implement various hyperparameter tuning methods with Hyperopt. To ensure that you can reproduce the code examples in this chapter, you will require the following:

- Python 3 (version 3.7 or above)
- The `pandas` package (version 1.3.4 or above)
- The `NumPy` package (version 1.21.2 or above)
- The `Matplotlib` package (version 3.5.0 or above)
- The `scikit-learn` package (version 1.0.1 or above)
- The `Hyperopt` package (version 0.2.7 or above)
- The `LightGBM` package (version 3.3.2 or above)

All the code examples for this chapter can be found on GitHub at `https://github.com/PacktPublishing/Hyperparameter-Tuning-with-Python`.

Introducing Hyperopt

All of the implemented optimization methods in the `Hyperopt` package assume we are working with a *minimization problem*. If your objective function is categorized as a maximization problem, for example, when you are using accuracy as the objective function score, you must *add a negative sign to your objective function*.

Utilizing the `Hyperopt` package to perform hyperparameter tuning is very simple. The following steps show how to perform any hyperparameter tuning methods provided in the `Hyperopt` package. More detailed steps, including the code implementation, will be given through various examples in the upcoming sections:

1. Define the objective function to be minimized.
2. Define the hyperparameter space.

3. (*Optional*) Initiate the `Trials()` object and pass it to the `fmin()` function.

4. Perform hyperparameter tuning by calling the `fmin()` function.

5. Train the model on full training data using the best set of hyperparameters that have been found from the output of the `fmin()` function.

6. Test the final trained model on the test data.

The simplest case of the objective function is when we only return the floating type of objective function score. However, we can also add other additional information to the output of the objective function, for example, the evaluation time or any other statistics we want to get for further analysis. When we add additional information to the output of the objective function score, `Hyperopt` expects the output of the objective function to be in the form of a Python dictionary that has at least two mandatory key-value pairs – that is, `status` and `loss`. The former key stores the status value of the run, while the latter key stores the objective function that we want to minimize.

The simplest type of hyperparameter space in Hyperopt is in the form of a Python dictionary, where the keys refer to the name of the hyperparameters and the values contain the distribution of the hyperparameters to be sampled from. The following example shows how we can define a very simple hyperparameter space in `Hyperopt`:

```
import numpy as np
from hyperopt import hp
hyperparameter_space = {
"criterion": hp.choice("criterion", ["gini", "entropy"]),
"n_estimators": 5 + hp.randint("n_estimators", 195),
"min_samples_split" : hp.loguniform("min_samples_split",
np.log(0.0001), np.log(0.5))
}
```

As you can see, the values of the `hyperparameter_space` dictionary are the distributions that accompany each of the hyperparameters we have in the space. `Hyperopt` provides a lot of sampling distributions that we can utilize, such as `hp.choice`, `hp.randint`, `hp.uniform`, `hp.loguniform`, `hp.normal`, and `hp.lognormal`. The `hp.choice` distribution will randomly choose one option from the several given options. The `hp.randint` distribution will randomly choose an integer within the range of `[0, high)`, where `high` is the input given by us. In the previous example, we passed `195` as the `high` value and added a value of 5. This means `Hyperopt` will randomly choose an integer within the range of `[5,200)`.

The rest of the distributions are dedicated to real/floating hyperparameter values. Note that Hyperopt also provides distributions dedicated to integer hyperparameter values that mimic the distribution of those four distributions – that is, `hp.quniform`, `hp.qloguniform`, `hp.qnormal`, and `hp.qlognormal`. For more information regarding the sampling distributions provided by Hyperopt, please refer to its official wiki page (`https://github.com/hyperopt/hyperopt/wiki/FMin#21-parameter-expressions`).

It is worth noting that Hyperopt enables us to define a **conditional hyperparameter space** (see *Chapter 4, Bayesian Optimization*) that suits our needs. The following code example shows how we can define such a search space:

```
hyperparameter_space =
hp.choice("class_weight_type", [
{"class_weight": None,
"n_estimators": 5 + hp.randint("none_n_estimators", 45),
},
{"class_weight": "balanced",
"n_estimators": 5 + hp.randint("balanced_n_estimators", 195),
}
])
```

As you can see, the only difference between a conditional hyperparameter space and a non-conditional one is that we add `hp.choice` before defining the hyperparameters for each condition. In this example, when `class_weight` is None, we will only search for the best `n_estimators` hyperparameters within the range [5,50]. On the other hand, when `class_weight` is "balanced", the range becomes [5,200].

Once the hyperparameter space is defined, we can start the hyperparameter tuning process via the `fmin()` function. The output of this function is the best set of hyperparameters that has been found from the tuning process. There are several important parameters available in this function that you need to know about. The `fn` parameter refers to the objective function we are trying to minimize, the `space` parameter refers to the hyperparameter space that will be used in our experiment, the `algo` parameter refers to the hyperparameter tuning algorithm that we want to utilize, the `rstate` parameter refers to the random seed for the tuning process, the `max_evals` parameter refers to the stopping criterion of the tuning process based on the number of trials, and the `timeout` parameter refers to the stopping criterion based on the time limit in seconds. Another important parameter is the `trials` parameter, which expects to receive the Hyperopt `Trials()` object.

The `Trials()` object in Hyperopt logs all the relevant information during the tuning process. This object is also responsible for storing all of the additional information we put in the dictionary output of the objective function. We can utilize this object for debugging purposes or to pass it directly to the built-in plotting module in Hyperopt.

Several built-in plotting modules are implemented in the `Hyperopt` package, such as `main_plot_history`, `main_plot_histogram`, and `main_plot_vars` modules. The first plotting module can help us understand the relationship between the loss values and the execution time. The second plotting module shows the histogram of all of the losses in all trials. The third plotting module is useful for understanding more about the heatmap of each hyperparameter in the space relative to the loss values.

Last but not least, it is worth noting that Hyperopt also supports parallel search processes by utilizing **MongoDB** or **Spark**. To utilize the parallel resources via MongoDB, we can simply change the trial database from `Trials()` to `MongoTrials()`. We can change from `Trials()` to `SparkTrials()` if we want to utilize Spark instead of MongoDB. Please refer to the official documentation of Hyperopt for more information about parallel computations (`https://github.com/hyperopt/hyperopt/wiki/Parallelizing-Evaluations-During-Search-via-MongoDB` and `http://hyperopt.github.io/hyperopt/scaleout/spark/`).

In this section, you were introduced to the overall capability of the `Hyperopt` package, along with the general steps to perform hyperparameter tuning with this package. In the next few sections, we will learn how to implement each of the hyperparameter tuning methods available in `Hyperopt` through examples.

Implementing Random Search

To implement Random Search (see *Chapter 3*) in Hyperopt, we can simply follow the steps explained in the previous section and pass the `rand.suggest` object to the `algo` parameter in the `fmin()` function. Let's learn how we can utilize the `Hyperopt` package to perform Random Search. We will use the same data and `sklearn` pipeline definition as in *Chapter 7, Hyperparameter Tuning via Scikit*, but with a slightly different definition of the hyperparameter space. Let's follow the steps that were introduced in the previous section:

1. Define the objective function to be minimized. Here, we are utilizing the defined pipeline, `pipe`, to calculate the *5-fold cross-validation* score by utilizing the `cross_val_score` function from `sklearn`. We will use the *F1 score* as the evaluation metric:

```python
import numpy as np
from sklearn.base import clone
from sklearn.model_selection import cross_val_score
from hyperopt import STATUS_OK
def objective(space):
    estimator_clone = clone(pipe).set_params(**space)
    return {'loss': -1 * np.mean(cross_val_
score(estimator_clone, X_train_full, y_train, cv=5,
scoring='f1', n_jobs=-1)),
            'status': STATUS_OK}
```

Note that the defined `objective` function only receives one input, which is the predefined hyperparameter space, `space`, and outputs a dictionary that contains two mandatory key-value pairs – that is, `status` and `loss`. It is also worth noting that the reason why we multiply the average cross-validation score output with `-1` is that `Hyperopt` always assumes that we are working with a minimization problem, while we are not in this example.

2. Define the hyperparameter space. Since we are using the `sklearn` pipeline as our estimator, we still need to follow the naming convention of the hyperparameters within the defined space (see *Chapter 7*). Note that the naming convention just needs to be applied to the hyperparameter names in the keys of the search space dictionary, not to the names within the sampling distribution objects:

```
from hyperopt import hp
hyperparameter_space = {
"model__n_estimators": 5 + hp.randint("n_estimators",
195),
"model__criterion": hp.choice("criterion", ["gini",
"entropy"]),
"model__class_weight": hp.choice("class_weight",
["balanced","balanced_subsample"]),
"model__min_samples_split": hp.loguniform("min_samples_
split", np.log(0.0001), np.log(0.5))
}
```

3. Initiate the `Trials()` object. In this example, we will utilize this object for plotting purposes after the tuning process has been done:

```
from hyperopt import Trials
trials = Trials()
```

4. Perform hyperparameter tuning by calling the `fmin()` function. Here, we are performing a Random Search by passing the defined objective function and hyperparameter space. We have set the `algo` parameter with the `rand.suggest` object and set the number of trials to `100` as the stopping criterion. We also set the random state to ensure reproducibility. Last but not least, we passed the defined `Trials()` object to the `trials` parameter:

```
from hyperopt import fmin, rand
best = fmin(objective,
            space=hyperparameter_space,
            algo=rand.suggest,
            max_evals=100,
            rstate=np.random.default_rng(0),
```

```
            trials=trials
        )
    print(best)
```

Based on the preceding code, we get around -0.621 of the objective function score, which refers to 0.621 of the average 5-fold cross-validation F--score. We also get a dictionary consisting of the best set of hyperparameters, as follows:

```
{'class_weight': 0, 'criterion': 1, 'min_samples_split':
0.00047017001935242104, 'n_estimators': 186}
```

As can be seen, Hyperopt will only return the index of the hyperparameter values when we use hp.choice as the sampling distribution (see the class_weight and criterion hyperparameters). Here, by referring to the predefined hyperparameter space, 0 for class_weight refers to *balanced* and 1 for criterion refers to *entropy*. Thus, the best set of hyperparameters is { 'model__class_weight': 'balanced', 'model__criterion': 'entropy', 'model__min_samples_split': 0.0004701700193524210, 'model__n_estimators': 186}.

5. Train the model on the full training data using the best set of hyperparameters that have been found in the output of the fmin() function:

```
pipe = pipe.set_params(**{'model__class_weight':
"balanced",
'model__criterion': "entropy",
'model__min_samples_split': 0.00047017001935242104,
'model__n_estimators': 186})
pipe.fit(X_train_full,y_train)
```

6. Test the final trained model on the test data:

```
from sklearn.metrics import f1_score
y_pred = pipe.predict(X_test_full)
print(f1_score(y_test, y_pred))
```

Based on the preceding code, we get around 0.624 for the F1-score when testing our final trained Random Forest model with the best set of hyperparameters on the test set.

7. Last but not least, we can also utilize the built-in plotting modules implemented in Hyperopt. The following code shows how to do this. Note that we need to pass the trials object from the tuning process to the plotting modules since all of the tuning process logs are in there:

```
from hyperopt import plotting
```

Now, we must plot the relationship between the loss values and the execution time:

```
plotting.main_plot_history(trials)
```

We will get the following output:

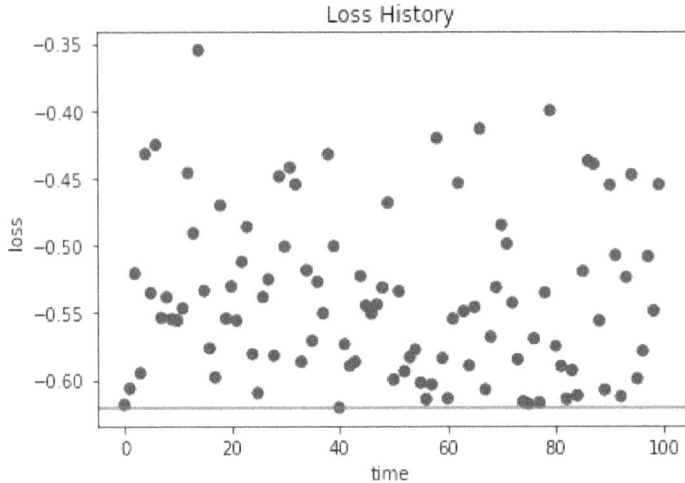

Figure 8.1 – Relationship between the loss values and the execution time

Now, we must plot the histogram of all of the objective function scores from all the trials:

```
plotting.main_plot_histogram(trials)
```

We will get the following output.

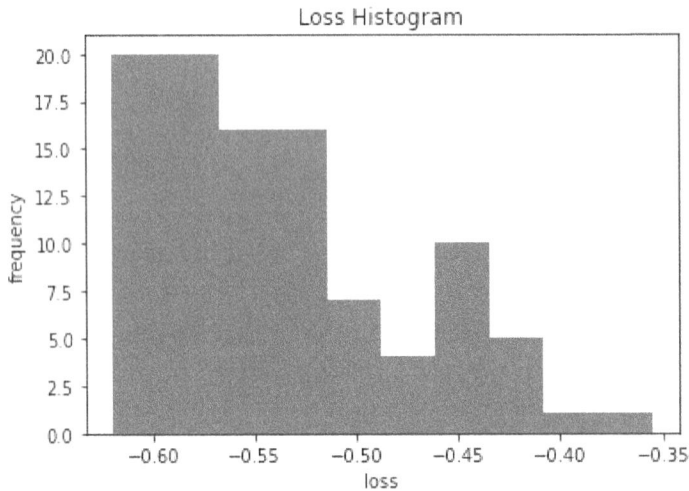

Figure 8.2 – Histogram of all of the objective function scores from all trials

Now, we must plot the heatmap of each hyperparameter in the space relative to the loss values:

```
Plotting.main_plot_vars(trials)
```

We will get the following output.

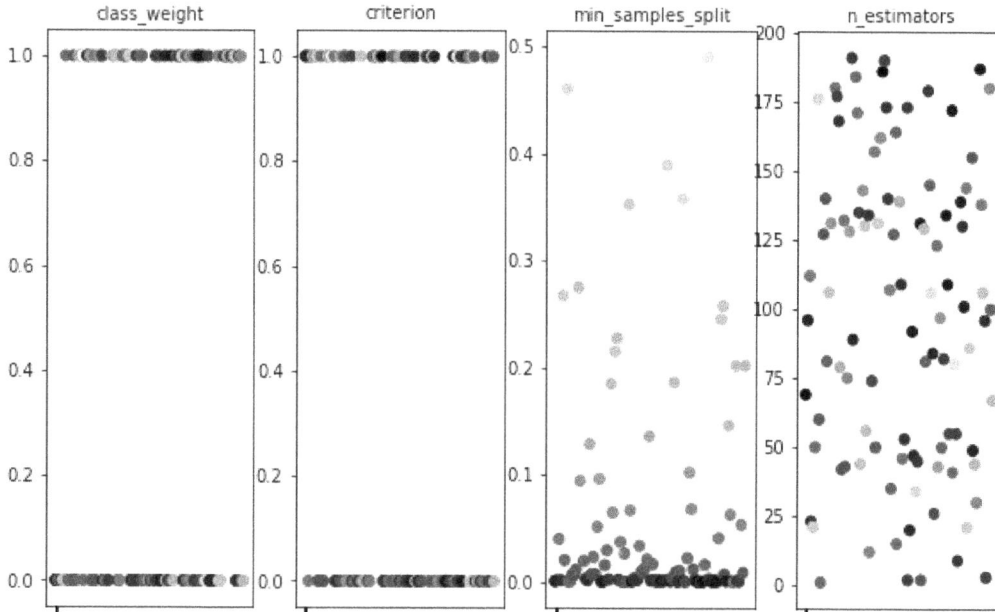

Figure 8.3 – Heatmap of each hyperparameter in the space
relative to the loss values (the darker, the better)

In this section, we learned how to perform Random Search in `Hyperopt` by looking at an example similar example to the one shown in *Chapter 7, Hyperparameter Tuning via Scikit*. We also saw what kind of figures we can get from utilizing the built-in plotting modules in Hyperopt.

It is worth noting that we are not bounded to using only the `sklearn` implementation of models to perform hyperparameter tuning with `Hyperopt`. We can also use implementations from other packages, such as `PyTorch`, `Tensorflow`, and so on. One thing that needs to be kept in mind is to be careful with the *data leakage issue* (see *Chapter 1, Evaluating Machine Learning Models*) when performing cross-validation. We must fit all of the data preprocessing methods on the training data and apply the fitted preprocessors to the validation data.

In the next section, we will learn how to utilize `Hyperopt` to perform hyperparameter tuning with one of the available Bayesian Optimization methods.

Implementing Tree-structured Parzen Estimators

Tree-Structured Parzen Estimators (**TPE**) is one of the variants of the Bayesian Optimization hyperparameter tuning group (see *Chapter 4, Exploring Bayesian Optimization*) that is also implemented in the `Hyperopt` package. To perform hyperparameter tuning with this method, we can follow a similar procedure as in the previous section by only changing the `algo` parameter to `tpe.suggest` in *Step 4*. The following code shows how to perform hyperparameter tuning with TPE in `Hyperopt`:

```
from hyperopt import fmin, tpe
best = fmin(objective,
            space=hyperparameter_space,
            algo=tpe.suggest,
            max_evals=100,
            rstate=np.random.default_rng(0),
            trials=trials
           )
print(best)
```

Using the same data, hyperparameter space, and parameters for the `fmin()` function, we get around `-0.620` for the objective function score, which refers to `0.620` of the average 5-fold cross-validation F1-score. We also get a dictionary consisting of the best set of hyperparameters, as follows:

```
{'class_weight': 1, 'criterion': 1, 'min_samples_split':
0.0005245304932726025, 'n_estimators': 138}
```

Once the model has been trained on the full data using the best set of hyperparameters, we get around `0.621` in terms of the F1-score when we test the final Random Forest model that's been trained on the test data.

In this section, we learned how to perform hyperparameter tuning using the TPE method with `Hyperopt`. In the next section, we will learn how to implement a variant of TPE called Adaptive TPE with the `Hyperopt` package.

Implementing Adaptive TPE

Adaptive TPE (**ATPE**) is a variant of the TPE hyperparameter tuning method that is developed based on several improvements compared to TPE, such as automatically tuning several hyperparameters of the TPE method based on the data that we have. For more information about this method, please refer to the original white papers. These can be found in the GitHub repository of the author (`https://github.com/electricbrainio/hypermax`).

While you can experiment with this method directly using the original GitHub repository of ATPE, `Hyperopt` has also included this method as part of the package. You can simply follow a similar procedure as in the *Implementing Random Search* section by only changing the `algo` parameter to `atpe.suggest` in *Step 4*. The following code shows how to perform hyperparameter tuning with ATPE in `Hyperopt`. Please note that ATPE utilizes the **LightGBM** model to predict each of the ATPE parameters. That's why we need to have the `LightGBM` package installed before we can start to perform hyperparameter tuning with ATPE in `Hyperopt`:

```
from hyperopt import fmin, atpe
best = fmin(objective,
            space=hyperparameter_space,
            algo=atpe.suggest,
            max_evals=100,
            rstate=np.random.default_rng(0),
            trials=trials
            )
print(best)
```

Using the same data, hyperparameter space, and parameters for the `fmin()` function, we get around `-0.621` for the objective function score, which refers to `0.621` of the average 5-fold cross-validation F1-score. We also get a dictionary consisting of the best set of hyperparameters, as follows:

```
{'class_weight': 1, 'criterion': 1, 'min_samples_split':
0.0005096354197481012, 'n_estimators': 157}
```

Once the model has been trained on the full data using the best set of hyperparameters, we get around `0.622` in terms of the F1 score when we test the final Random Forest model that was trained on the test data.

In this section, we learned how to perform hyperparameter tuning using the ATPE method with `Hyperopt`. In the next section, we will learn how to implement a hyperparameter tuning method that is part of the Heuristic Search group with the `Hyperopt` package.

Implementing simulated annealing

Simulated annealing (SA) is part of the Heuristic Search hyperparameter tuning group (see *Chapter 5, Exploring Heuristic Search*), which is also implemented in the Hyperopt package. Similar to TPE and ATPE, to perform hyperparameter tuning with this method, we can simply follow the procedure shown in the *Implementing Random Search* section; we only need to change the algo parameter to anneal.suggest in *Step 4*. The following code shows how to perform hyperparameter tuning with SA in Hyperopt:

```
from hyperopt import fmin, anneal
best = fmin(objective,
            space=hyperparameter_space,
            algo=anneal.suggest,
            max_evals=100,
            rstate=np.random.default_rng(0),
            trials=trials
            )
print(best)
```

Using the same data, hyperparameter space, and parameters for the fmin() function, we get around -0.620 for the objective function score, which refers to 0.620 of the average 5-fold cross-validation F1-score. We also get a dictionary consisting of the best set of hyperparameters, as follows:

```
{'class_weight': 1, 'criterion': 1, 'min_samples_split':
0.00046660708302994583, 'n_estimators': 189}
```

Once the model has been trained on the full data using the best set of hyperparameters, we get around 0.625 in terms of the F1-score when we test the final Random Forest model that was trained on the test data.

While Hyperopt has built-in plotting modules, we can also create a customized plotting function by utilizing the Trials() object. The following code shows how to visualize the distribution of each hyperparameter over the number of trials:

1. Get the value of each hyperparameter in each of the trials:

```
plotting_data = np.array([[x['result']['loss'],
x['misc']['vals']['class_weight'][0],
x['misc']['vals']['criterion'][0],
x['misc']['vals']['min_samples_split'][0],
x['misc']['vals']['n_estimators'][0],
] for x in trials.trials])
```

2. Convert the values into a pandas DataFrame:

```
import pandas as pd
plotting_data = pd.DataFrame(plotting_data,
columns=['score', 'class_weight', 'criterion', 'min_
samples_split','n_estimators'])
```

3. Plot the relationship between each hyperparameter's distribution and the number of trials:

```
import matplotlib.pyplot as plt
plotting_data.plot(subplots=True,figsize=(12, 12))
plt.xlabel("Iterations")
plt.show()
```

Based on the preceding code, we will get the following output:

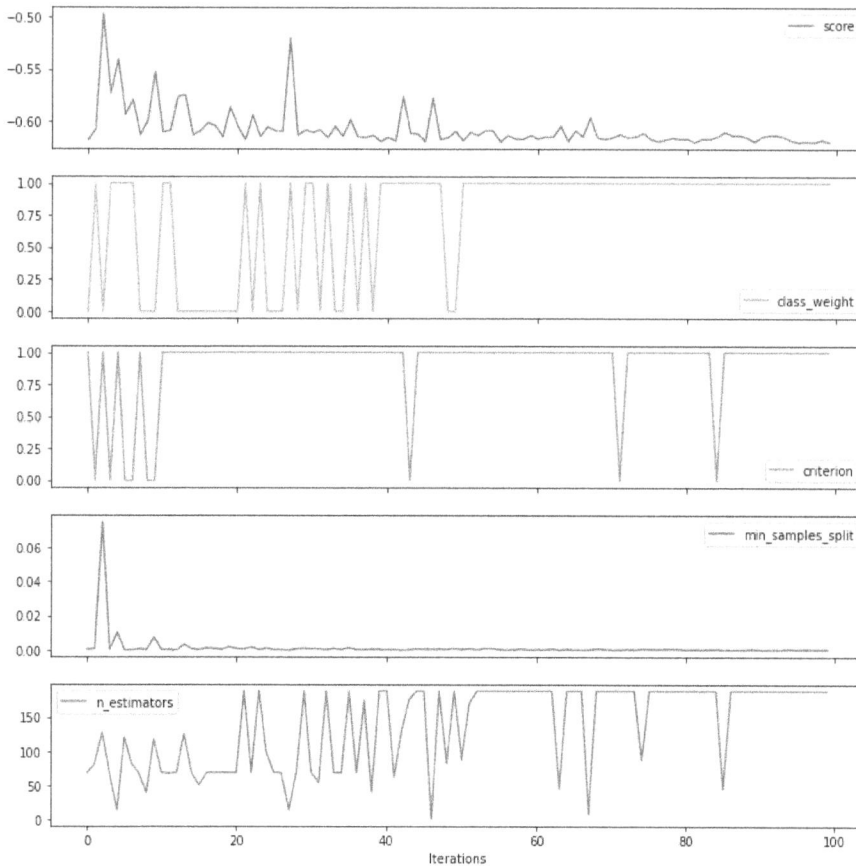

Figure 8.4 – Relationship between each hyperparameter's distribution and the number of trials

In this section, we learned how to implement SA in `Hyperopt` by using the same example as in the *Implementing Random Search* section. We also learned how to create a custom plotting function to visualize the relationship between each hyperparameter's distribution and the number of trials.

Summary

In this chapter, we learned all the important things about the `Hyperopt` package, including its capabilities and limitations, and how to utilize it to perform hyperparameter tuning. We saw that `Hyperopt` supports various types of sampling distribution methods but can only work with a minimization problem. We also learned how to implement various hyperparameter tuning methods with the help of this package, which has helped us understand each of the important parameters of the classes and how are they related to the theory that we learned about in the previous chapters. At this point, you should be able to utilize `Hyperopt` to implement your chosen hyperparameter tuning method and, ultimately, boost the performance of your ML model. Equipped with the knowledge from *Chapter 3*, to *Chapter 6*, you should be able to understand what's happening if there are errors or unexpected results, as well as understand how to set up the method configuration so that it matches your specific problem.

In the next chapter, we will learn about the `Optuna` package and how to utilize it to perform various hyperparameter tuning methods. The goal of the next chapter is similar to this chapter – that is, being able to utilize the package for hyperparameter tuning purposes and understanding each of the parameters of the implemented classes.

9

Hyperparameter Tuning via Optuna

Optuna is a Python package that provides various implementations of hyperparameter tuning methods, including but not limited to Grid Search, Random Search, and **Tree-Structured Parzen Estimators (TPE)**. This package also enables us to create our own hyperparameter tuning method class and integrate it with other popular hyperparameter tuning packages, such as `scikit-optimize`.

In this chapter, you'll be introduced to the `Optuna` package, starting with its numerous features, how to utilize it to perform hyperparameter tuning, and all of the other important things you need to know about `Optuna`. We'll not only learn how to utilize `Optuna` to perform hyperparameter tuning with their default configurations but also discuss the available configurations along with their usage. Moreover, we'll also discuss how the implementation of the hyperparameter tuning methods is related to the theory that we have learned in previous chapters, since there may be some minor differences or adjustments made in the implementation.

By the end of this chapter, you will be able to understand all of the important things you need to know about `Optuna` and implement various hyperparameter tuning methods available in this package. You'll also be able to understand each of the important parameters of the classes and how they are related to the theory that we have learned in previous chapters. Finally, equipped with the knowledge from previous chapters, you will also be able to understand what's happening if there are errors or unexpected results and understand how to set up the method configuration to match your specific problem.

The following are the main topics that will be discussed in this chapter:

- Introducing Optuna
- Implementing TPE
- Implementing Random Search
- Implementing Grid Search
- Implementing Simulated Annealing
- Implementing Successive Halving
- Implementing Hyperband

Technical requirements

We will learn how to implement various hyperparameter tuning methods with `Optuna`. To ensure that you are able to reproduce the code examples in this chapter, you will require the following:

- Python 3 (version 3.7 or above)
- Installed `pandas` package (version 1.3.4 or above)
- Installed `NumPy` package (version 1.21.2 or above)
- Installed `Matplotlib` package (version 3.5.0 or above)
- Installed `scikit-learn` package (version 1.0.1 or above)
- Installed `Tensorflow` package (version 2.4.1 or above)
- Installed `Optuna` package (version 2.10.0 or above)

All of the code examples for this chapter can be found on GitHub at `https://github.com/PacktPublishing/Hyperparameter-Tuning-with-Python`.

Introducing Optuna

`Optuna` is a hyperparameter tuning package in Python that provides several hyperparameter tuning methods implementation, such as Grid Search, Random Search, Tree-structured Parzen Estimators (TPE), and many more. Unlike `Hyperopt`, which assumes we are always working with a minimization problem (see *Chapter 8, Hyperparameter Tuning via Hyperopt*), we can tell `Optuna` the type of optimization problem we are working on: minimization or maximization.

Optuna has two main classes, namely **samplers** and **pruners**. Samplers are responsible for performing the hyperparameter tuning optimization, whereas pruners are responsible for judging whether we should prune the trials based on the reported values. In other words, pruners act like *early stopping methods* where we will stop a hyperparameter tuning iteration whenever it seems that there's no additional benefit to continuing the process.

The built-in implementation for samplers includes several hyperparameter tuning methods that we have learned in *Chapters 3 - 4*, namely Grid Search, Random Search, and TPE, and also other methods that are outside of the scope of this book, such as CMA-ES, NSGA-II, and many more. We can also define our own custom samplers, such as the Simulated Annealing (SA), which will be discussed in the upcoming section. Furthermore, Optuna also allows us to integrate samplers from another package, such as from the scikit-optimize (skopt) package where we can utilize many Bayesian optimization-based methods from there.

> **Integrations in Optuna**
>
> Besides skopt, there are also many other integrations provided by Optuna, including but not limited to scikit-learn, Keras, PyTorch, XGBoost, LightGBM, FastAI, MLflow, and many more. For more information about the available integrations, please see the official documentation (https://optuna.readthedocs.io/en/v2.10.0/reference/integration.html).

As for pruners, Optuna provides both statistics-based and multi-fidelity optimization (MFO)-based methods. There are MedianPruner, PercentilePruner, and ThresholdPruner for the statistics-based group. MedianPruner will prune the trials whenever the current trial's best intermediate result is worse compared to the median of the result of the previous trial. PercentilePruner will perform pruning when the current best intermediate value is part of the bottom percentile from previous trials. ThresholdPruner will simply perform pruning whenever the predefined threshold is met. The MFO-based pruners implemented in Optuna are SuccessiveHalvingPruner and HyperbandPruner. Both of them *define the resource as the number of training steps or epochs*, not as the number of samples such as in the implementations of scikit-learn. We will learn how to utilize these MFO-based pruners in the upcoming sections.

To perform hyperparameter tuning with Optuna, we can simply perform the following simple steps (more detailed steps, including the code implementation, will be given through various examples in the upcoming sections):

1. Define the objective function along with the hyperparameter space.
2. Initiate a study object via the create_study() function.
3. Perform hyperparameter tuning by calling the optimize() method on the study object.
4. Train the model on full training data using the best set of hyperparameters found.
5. Test the final trained model on the test data.

In Optuna, we can directly define the hyperparameter space within the objective function itself. There's no need to define another dedicated separate object just to store the hyperparameter space. This means that implementing conditional hyperparameters in Optuna becomes very easy since we just need to put them within the corresponding if-else blocks in the objective function. Optuna also provides very handy hyperparameter sampling distribution methods including suggest_categorical, suggest_discrete_uniform, suggest_int, and suggest_float.

The suggest_categorical method will suggest value from a categorical type of hyperparameters, which works similarly with the random.choice() method. The suggest_discrete_uniform can be utilized for a discrete type of hyperparameters, which works very similar to the hp.quniform in Hyperopt (see *Chapter 8, Hyperparameter Tuning via Hyperopt*) by sampling uniformly from the range of [low, high] with a q step of discretization. The suggest_int method works similarly to the random.randint() method. Finally, the suggest_float method. This method works for a floating type of hyperparameters and is actually a wrapper of two other sampling distribution methods, namely the suggest_uniform and suggest_loguniform. To utilize suggest_loguniform, simply set the log parameter in suggest_float as True.

To have a better understanding of how we can define the hyperparameter space within the objective function, the following code shows an example of how to define an objective function using **TFKeras**. Note that in this example, we write several functions to be called in the objective function, to ensure readability and to enable us to write the code in a modular fashion. However, you can also put all of the code within one single objective function directly. The data and preprocessing steps used in this example are the same as in *Chapter 7, Hyperparameter Tuning via Scikit*. However, in this example, we are using a **neural network** model instead of a random forest as follows:

1. Create a function to define the model architecture. Here, we create a binary classifier model where the number of hidden layers, number of units, dropout rate, and the activation function for each layer are part of the hyperparameter space, as follows:

```
import optuna
from tensorflow.keras.models import Sequential
from tensorflow.keras.layers import Dense, Dropout
def create_model(trial: optuna.trial.Trial, input_size:
int):
model = Sequential()
model.add(Dense(input_size,input_shape=(input_
size,),activation='relu'))
 num_layers = trial.suggest_int('num_
layers',low=0,high=3)
for layer_i in range(num_layers):
n_units = trial.suggest_int(f'n_units_layer_
```

```
{layer_i}',low=10,high=100,step=5)
 dropout_rate = trial.suggest_float(f'dropout_rate_layer_
{layer_i}',low=0,high=0.5)
actv_func = trial.suggest_categorical(f'actv_func _layer_
{layer_i}',['relu','tanh','elu'])
model.add(Dropout(dropout_rate))
 model.add(Dense(n_units,activation=actv_func))
model.add(Dense(1,activation='sigmoid'))
return model
```

2. Create a function to define the model's optimizer. Notice that we define conditional hyperparameters in this function where we have a different set of hyperparameters for a different chosen optimizer as follows:

```
import tensorflow as tf
def create_optimizer(trial: optuna.trial.Trial):
opt_kwargs = {}
opt_selected = trial.suggest_categorical('optimizer',
['Adam','SGD'])
if opt_selected == 'SGD':
opt_kwargs['lr'] = trial.suggest_float('sgd_lr',1e-5,1e-
1,log=True)
opt_kwargs['momentum'] = trial.suggest_float('sgd_
momentum',1e-5,1e-1,log=True)
else: #'Adam'
opt_kwargs['lr'] = trial.suggest_float('adam_lr',1e-5,1e-
1,log=True)
optimizer = getattr(tf.optimizers,opt_selected)(**opt_
kwargs)
return optimizer
```

3. Create the `train` and `validation` functions. Note that the preprocessing code is not shown here, but you can see it in the GitHub repo mentioned in the *Technical requirements* section for the full code. As the case with the examples in *Chapter 7*, we are also using F1-score as the evaluation metric of the model as follows:

```
def train(trial, df_train: pd.DataFrame, df_val:
pd.DataFrame = None):
    X_train,y_train = df_train.drop(columns=['y']), df_
train['y']
    if df_val is not None:
```

```
        X_val,y_val = df_val.drop(columns=['y']), df_
val['y']
    #Apply pre-processing here...
    #...
    #Build model & optimizer
    model = create_model(trial,X_train.shape[1])
    optimizer = create_optimizer(trial)
    model.compile(loss='binary_crossentropy',
optimizer=optimizer, metrics=[f1_m])
    history = model.fit(X_train,y_train,
                    epochs=trial.suggest_
int('epoch',15,50),
                batch_size=64,
                validation_data=(X_val,y_val) if df_val is
not None else None)
    if df_val is not None:
        return np.mean(history.history['val_f1_m'])
    else:
        return model
```

4. Create the `objective` function. Here, we split the original training data into training data for hyperparameter tuning, `df_train_hp`, and validation data, `df_val`. We won't follow the k-fold cross-validation evaluation method since it will take too much time for the neural network model to go through several folds of evaluation within each tuning trial (see *Chapter 1, Evaluating Machine Learning Models*).

```
from sklearn.model_selection import train_test_split
def objective(trial: optuna.trial.Trial, df_train:
pd.DataFrame):
#Split into Train and Validation data
    df_train_hp, df_val = train_test_split(df_train,
test_size=0.1, random_state=0)
    #Train and Validate Model
    val_f1_score = train(trial, df_train_hp, df_val)
    return val_f1_score
```

To perform hyperparameter tuning in `Optuna`, we need to initiate a `study` object via the `create_study()` function. The `study` object provides interfaces to run a new `Trial` object and access the trials' history. The `Trial` object is simply an object that involves the process of evaluating an `objective` function. This object will be passed to the `objective` function and it is responsible for managing the trial's state, providing interfaces upon receiving the parameter suggestion just as we saw earlier in the `objective` function. The following code shows how to utilize the `create_study()` function to initiate a `study` object:

```
study = optuna.create_study(direction='maximize')
```

There are several important parameters in the `create_study()` function. The `direction` parameter allows us to tell `Optuna` what kind of optimization problem we are working on. There are two valid values for this parameter, namely *maximize* and *minimize*. By setting the `direction` parameter equal to *maximize*, it means that we tell `Optuna` that we are currently working on a maximization problem. `Optuna` sets this parameter to *minimize* by default. The `sampler` parameter refers to the hyperparameter tuning algorithm that we want to use. By default, `Optuna` will use TPE as the sampler. The `pruner` parameter refers to the pruning algorithm that we want to use, where `MedianPruner()` is used by default.

> **Pruning in Optuna**
>
> Although `MedianPruner()` is chosen by default, the pruning process will not be performed unless we explicitly tell `Optuna` to do so within the `objective` function. This example shows how to perform a simple pruning procedure with the default pruner in `Optuna` at the following link: `https://github.com/optuna/optuna-examples/blob/main/simple_pruning.py`.

Besides the three preceding parameters, there are also other parameters in the `create_study()` function, namely `storage`, `study_name`, and `load_if_exists`. The `storage` parameter expects a database URL input, which will be handled with **SQLAlchemy** internally by `Optuna`. If we do not pass a database URL, `Optuna` will use the in-memory storage instead. The `study_name` parameter is simply the name that we want to give to the current `study` object. If we do not pass a name, `Optuna` will automatically generate a random name for us. Last but not least, the `load_if_exists` parameter is a Boolean parameter that handles cases when there might be conflicting study names. If the study name is already generated in the storage, and we set `load_if_exists=False`, then `Optuna` will raise an error. On the other hand, if the study name is already generated in the storage, but we set `load_if_exists=True`, `Optuna` will just load the existing `study` object instead of creating a new one.

Once the `study` object is initiated along with the appropriate parameters, we can start performing the hyperparameter tuning by calling the `optimize()` method. The following code shows you how to do that:

```
study.optimize(func=lambda trial: objective(trial, df_train),
               n_trials=50, n_jobs=-1)
```

There are several important parameters in the `optimize()` method. The first and most important method is the `func` parameter. This parameter expects a callable that implements the `objective` function. Here, we don't directly pass the `objective` function to the `func` parameter since our `objective` function expects two inputs, while by default, Optuna can only handle an `objective` function with one input, which is the `Trial` object itself. That's why we need the help of Python's built-in `lambda` function to pass the second input to our `objective` function. You can also utilize the same `lambda` function if your `objective` function has more than two inputs.

The second most important parameter is `n_trials`, which refers to the number of trials or iterations for the hyperparameter tuning process. Another implemented parameter that can be used as the stopping criteria is the `timeout` parameter. This parameter expects the stopping criteria in the unit of seconds. By default, Optuna sets the `n_trials` and `timeout` parameters to None. If we leave it be, then Optuna will run the hyperparameter tuning process until it receives a termination signal, such as `Ctrl+C` or `SIGTERM`.

Last but not least, Optuna also allows us to utilize the parallel resources through a parameter called `n_jobs`. By default, Optuna will set `n_jobs=1`, meaning that it will only utilize one job. Here, we set `n_jobs=-1`, meaning that we will use all of the CPU counts in our computer to perform parallel computation.

> **Hyperparameter's Importance in Optuna**
>
> Optuna provides a very nice module to measure the importance of each hyperparameter in the search space. As per version 2.10.0, there are two methods implemented, namely the **fANOVA** and **Mean Decrease Impurity** methods. Please see the official documentation on how to utilize this module and the theory behind the implemented methods, available at the following link: `https://optuna.readthedocs.io/en/v2.10.0/reference/importance.html`.

In this section, we learned what Optuna is in general, the available features that we can utilize, and the general steps as to how to perform hyperparameter tuning with this package. Optuna also has various visualization modules that can help us track our hyperparameter tuning experiments, which will be discussed in *Chapter 13, Tracking Hyperparameter Tuning Experiments*. In the upcoming sections, we will learn how to perform various hyperparameter tuning methods with Optuna through examples.

Implementing TPE

TPE is one of the variants of the Bayesian optimization hyperparameter tuning group (see *Chapter 4*), which is the default sampler in Optuna. To perform hyperparameter tuning with TPE in Optuna, we can just simply pass the optuna.samplers.TPESampler() class to the sampler parameter of the create_study() function. The following example shows how to implement TPE in Optuna. We'll use the same data as in the examples in *Chapter 7* and follow the steps introduced in the preceding section as follows:

1. Define the objective function along with the hyperparameter space. Here, we'll use the same function that we defined in the *Introducing Optuna* section. Remember that we use the train-validation split instead of the k-fold cross-validation method within the objective function.

2. Initiate a study object via the create_study() function as follows:

    ```
    study = optuna.create_study(direction='maximize',
        sampler=optuna.samplers.TPESampler(seed=0))
    ```

3. Perform hyperparameter tuning by calling the optimize() method on the study object as follows:

    ```
    study.optimize(lambda trial: objective(trial, df_train),
                    n_trials=50, n_jobs=-1)
    print("Best Trial:")
    best_trial = study.best_trial
    print("    Value: ", best_trial.value)
    print("    Hyperparameters: ")
    for key, value in best_trial.params.items():
        print(f"        {key}: {value}")
    ```

Based on the preceding code, we get around 0.563 of F1-score evaluated in the validation data. We also get a dictionary consisting of the best set of hyperparameters as follows:

```
{'num_layers': 2,'n_units_layer_0': 30,'dropout_
rate_layer_0': 0.14068484717257745,'actv_
func_layer_0': 'relu','n_units_layer_1':
20,'dropout_rate_layer_1': 0.34708586671782293,'actv_
func_layer_1': 'relu','optimizer': 'Adam','adam_lr':
0.0018287924415952158,'epoch': 41}
```

4. Train the model on full training data using the best set of hyperparameters found. Here, we define another function called `train_and_evaluate_final()` that has the purpose of training the model in the full training data based on the best set of hyperparameters found in the preceding step, as well as evaluating it on the test data. You can see the implemented function in the GitHub repo mentioned in the *Technical requirements* section. Define the function as follows:

    ```
    train_and_evaluate_final(df_train, df_test, **best_trial.
    params)
    ```

5. Test the final trained model on the test data. Based on the results from the preceding step, we get around `0.604` in F1-score when testing our final trained neural network model with the best set of hyperparameters on the test set.

There are several important parameters for the `TPESampler` class. First, there is the `gamma` parameter, which refers to the threshold used in TPE to divide good and bad samples (see *Chapter 4*). The `n_startup_trials` parameter is responsible for controlling how many trials will utilize Random Search before starting to perform the TPE algorithm. The `n_ei_candidates` parameter is responsible for controlling how many candidate samples are used to calculate the `expected improvement acquisition` function. Last but not least, the `seed` parameter, which controls the random seed of the experiment. There are many other parameters available for the `TPESampler` class, so please see the original documentation for more information, available at the following link: `https://optuna.readthedocs.io/en/v2.10.0/reference/generated/optuna.samplers.TPESampler.html`.

In this section, we have learned how to perform hyperparameter tuning with TPE in `Optuna` using the same data as in the example in *Chapter 7*. As mentioned in *Chapter 4, Exploring Bayesian Optimization* `Optuna` also implements the multivariate TPE, which is able to capture the interdependencies among hyperparameters. To enable the multivariate TPE, we can just simply set the `multivariate` parameter in `optuna.samplers.TPESampler()` as `True`. In the next section, we will learn how to perform Random Search with `Optuna`.

Implementing Random Search

Implementing Random Search in `Optuna` is very similar to implementing TPE in `Optuna`. We can just follow a similar procedure to the preceding section and change the `sampler` parameter in the `optimize()` method in *step 2*. The following code shows you how to do that:

```
study = optuna.create_study(direction='maximize',
sampler=optuna.samplers.RandomSampler(seed=0))
```

Using the exact same data, preprocessing steps, hyperparameter space, and `objective` function, we get around `0.548` in the F1-score evaluated in the validation data. We also get a dictionary consisting of the best set of hyperparameters as follows:

```
{'num_layers': 0,'optimizer': 'Adam','adam_lr':
0.05075826567070766,'epoch': 50}
```

After the model is trained with full data using the best set of hyperparameters, we get around `0.596` in F1-score when we test the final neural network model trained on the test data. Notice that although we have defined many hyperparameters earlier, (see the `objective` function in the preceding section), here, we do not get all of them in the results. This is because most of the hyperparameters are conditional hyperparameters. For example, since the chosen value for the *'num_layers'* hyperparameter is zero, there will be no *'n_units_layer_{layer_i}'*, *'dropout_rate_layer_{layer_i}'*, or *'actv_func_layer_{layer_i}'* since those hyperparameters will only exist when the *'num_layers'* hyperparameter is greater than zero.

In this section, we have seen how to perform hyperparameter tuning using the Random Search method with `Optuna`. In the next section, we will learn how to implement Grid Search with the `Optuna` package.

Implementing Grid Search

Implementing Grid Search in `Optuna` is a bit different from implementing TPE and Random Search. Here, we need to also define the search space object and pass it to `optuna.samplers.GridSampler()`. The search space object is just a Python dictionary data structure consisting of hyperparameters' names as the keys and the possible values of the corresponding hyperparameter as the dictionary's values. `GridSampler` will stop the hyperparameter tuning process if all of the combinations in the search space have already been evaluated, even though the number of trials, `n_trials`, passed to the `optimize()` method has not been reached yet. Furthermore, `GridSampler` will only get the value stated in the search space no matter the range we pass to the sampling distribution methods, such as `suggest_categorical`, `suggest_discrete_uniform`, `suggest_int`, and `suggest_float`.

The following code shows how to perform Grid Search in `Optuna`. The overall procedure to implement Grid Search in `Optuna` is similar to the procedure stated in the *Implementing Tree-structured Parzen Estimators* section. The only difference is that we have to define the search space and change the `sampler` parameter to `optuna.samplers.GridSampler()` in the `optimize()` method in *step 2* as follows:

```
search_space = {'num_layers': [0,1],
                'n_units_layer_0': list(range(10,50,5)),
                'dropout_rate_layer_0': np.linspace(0,0.5,5),
                'actv_func_layer_0': ['relu','elu'],
```

```
                    'optimizer': ['Adam','SGD'],
                    'sgd_lr': np.linspace(1e-5,1e-1,5),
                    'sgd_momentum': np.linspace(1e-5,1e-1,5),
                    'adam_lr': np.linspace(1e-5,1e-1,5),
                    'epoch': list(range(15,50,5))
                }
    study = optuna.create_study(direction='maximize',
    sampler=optuna.samplers.GridSampler(search_space),
                            )
```

Based on the preceding code, we get around 0.574 of the F1-score evaluated in the validation data. We also get a dictionary consisting of the best set of hyperparameters as follows:

```
{'num_layers': 0,'optimizer': 'Adam','adam_lr':
0.05000500000000001,'epoch': 25}
```

After the model is trained on full data using the best set of hyperparameters, we get around 0.610 in F1-score when we test the final neural network model trained on the test data.

It is worth noting that GridSampler will rely on the search space to perform the hyperparameter sampling. For example, in the search space, we only define the valid values for num_layers as [0,1]. So, although within the objective function we set trial.suggest_int('num_layers',low=0,high=3) (see the *Introducing Optuna* section), only 0 and 1 will be tested during the tuning process. Remember that, in Optuna, we can specify the stopping criterion through the n_trials or timeout parameters. If we specify either one of those criteria, GridSampler will not test all of the possible combinations in the search space; the tuning process will stop once the stopping criterion is met. In this example, we set n_trials=50, just like the example in the preceding section.

In this section, we have learned how to perform hyperparameter tuning using the Grid Search method with Optuna. In the next section, we will learn how to implement SA with the Optuna package.

Implementing Simulated Annealing

SA is not part of the built-in implementation of the hyperparameter tuning method in Optuna. However, as mentioned in the first section of this chapter, we can define our own custom sampler in Optuna. When creating a custom sampler, we need to create a class that inherits from the BaseSampler class. The most important method that we need to define within our custom class is the sample_relative() method. This method is responsible for sampling the corresponding hyperparameters from the search space based on the hyperparameter tuning algorithm we chose.

The complete custom `SimulatedAnnealingSampler()` class with geometric cooling annealing schedule (see *Chapter 5*) has been defined and can be seen in the GitHub repo mentioned in the *Technical requirements* section. The following code shows only the implementation of the `sample_relative()` method within the class:

```
class SimulatedAnnealingSampler(optuna.samplers.BaseSampler):
    ...
    def sample_relative(self, study, trial, search_space):
        if search_space == {}:
            # The relative search space is empty (it means this
is the first trial of a study).
            return {}

        prev_trial = self._get_last_complete_trial(study)
        if self._rng.uniform(0, 1) <= self._transition_
probability(study, prev_trial):
            self._current_trial = prev_trial
        params = self._sample_neighbor_params(search_space)
        #Geometric Cooling Annealing Schedule
        self._temperature *= self.cooldown_factor
        return params
    ...
```

The following code shows how to perform hyperparameter tuning with SA in `Optuna`. The overall procedure to implement SA in `Optuna` is similar to the procedure stated in the *Implementing Tree-structured Parzen Estimators* section. The only difference is that we have to change the `sampler` parameter to `SimulatedAnnealingSampler()` in the `optimize()` method in *step 2* as follows:

```
study = optuna.create_study(direction='maximize',
                    sampler=SimulatedAnnealingSampler(seed=0),
                    )
```

Using the exact same data, preprocessing steps, hyperparameter space, and `objective` function, we get around `0.556` of the F1-score evaluated in the validation data. We also get a dictionary consisting of the best set of hyperparameters as follows:

```
{'num_layers': 3,'n_units_layer_0': 30,'dropout_rate_
layer_0': 0.28421697443432425,'actv_func_layer_0':
'tanh','n_units_layer_1': 20,'dropout_rate_layer_1':
0.05936385947712203,'actv_func_layer_1': 'tanh','n_units_
layer_2': 25,'dropout_rate_layer_2': 0.2179324626328134,'actv_
```

```
func_layer_2': 'relu','optimizer': 'Adam','adam_lr':
0.006100619734336806,'epoch': 39}
```

After the model is trained on full data using the best set of hyperparameters, we get around 0.559 in F1-score when we test the final neural network model trained on the test data.

In this section, we have learned how to perform hyperparameter tuning using the SA algorithm with Optuna. In the next section, we will learn how to utilize Successive Halving as a pruning method in Optuna.

Implementing Successive Halving

Successive Halving (SH) is implemented as a pruner in Optuna, meaning that it is responsible for stopping hyperparameter tuning iterations whenever it seems that there's no additional benefit to continuing the process. Since it is implemented as a pruner, the resource definition of SH (see *Chapter 6*) in Optuna refers to the number of training steps or epochs of the model, instead of the number of samples, as it does in scikit-learn's implementation.

We can utilize SH as a pruner along with any sampler that we use. This example shows you how to perform hyperparameter tuning with the Random Search algorithm as the sampler and SH as the pruner. The overall procedure is similar to the procedure stated in the *Implementing TPE* section. Since we are utilizing SH as a pruner, we have to edit our objective function so that it will utilize the pruner during the optimization process. In this example, we can use the callback integration with TFKeras provided by Optuna via optuna.integration.TFKerasPruningCallback. We simply need to pass this class to the callbacks parameter when fitting the model within the train function as shown in the following code:

```
def train(trial, df_train: pd.DataFrame, df_val: pd.DataFrame =
None):
...

    history = model.fit(X_train,y_train,
                        epochs=trial.suggest_int('epoch',15,50),
                        batch_size=64,
                        validation_data=(X_val,y_val) if df_val
is not None else None,
                        callbacks=[optuna.integration.
TFKerasPruningCallback(trial,'val_f1_m')],
                    )

...
```

Once we have told `Optuna` to utilize the pruner, we also need to set the `pruner` parameter in the `optimize()` method to `optuna.pruners.SuccessiveHalvingPruner()` in *step 2* of the *Implementing Tree-structured Parzen Estimators* section as follows:

```
study = optuna.create_study(direction='maximize',
    sampler=optuna.samplers.RandomSampler(seed=0),
    pruner=optuna.pruners.SuccessiveHalvingPruner(reduction_
factor=3, min_resource=5)
                                )
```

In this example, we also increased the number of trials from 50 to 100 since most of the trials will be pruned anyway as follows:

```
study.optimize(lambda trial: objective(trial, df_train),
                n_trials=100, n_jobs=-1,
                )
```

Using the exact same data, preprocessing steps, and hyperparameter space, we get around 0.582 of the F1-score evaluated in the validation data. Out of 100 trials performed, there are 87 trials pruned by SH, which implies only 13 completed trials. We also get a dictionary consisting of the best set of hyperparameters as follows:

```
{'num_layers': 3,'n_units_layer_0': 10,'dropout_rate_
layer_0': 0.03540368984067649,'actv_func_layer_0':
'elu','n_units_layer_1': 15,'dropout_rate_layer_1':
0.008554081181978979,'actv_func_layer_1': 'elu','n_units_
layer_2': 15,'dropout_rate_layer_2': 0.4887044768096681,'actv_
func_layer_2': 'relu','optimizer': 'Adam','adam_lr':
0.02763126523504823,'epoch': 28}
```

After the model is trained on full data using the best set of hyperparameters, we get around 0.597 in F1-score when we test the final neural network model trained on the test data.

It is worth noting that there are several parameters for `SuccessiveHalvingPruner` that we can customize based on our needs. The `reduction_factor` parameter refers to the multiplier factor of SH (see *Chapter 6*). The `min_resource` parameter refers to the minimum number of resources to be used at the first trial. This parameter is set to *'auto'*, by default, where a heuristic is utilized to calculate the most appropriate value based on the number of required steps for the first trial to be completed. In other words, `Optuna` will only be able to start the tuning process after the `min_resource` training steps or epochs have been performed.

Optuna also provides the `min_early_stopping_rate` parameter, which has the exact same meaning as we defined in *Chapter 6*. Last but not least, the `bootstrap_count` parameter. This parameter is not part of the original SH algorithm. The purpose of this parameter is to control the minimum number of trials that need to be completed before the actual SH iterations start.

You may wonder, what about the parameter that controls the value of maximum resources and the number of candidates in SH? Here, in `Optuna`, the maximum resources definition will be automatically derived based on the total number of training steps or epochs within the defined `objective` function. As for the parameter that controls the number of candidates, `Optuna` delegates this responsibility to the `n_trials` parameter in the `study.optimize()` method.

In this section, we have learned how to utilize SH as a pruner during the hyperparameter tuning process. In the next section, we will learn how to utilize Hyperband, the extended algorithm of SH, as a pruning method in `Optuna`.

Implementing Hyperband

Implementing **Hyperband** (**HB**) in `Optuna` is very similar to implementing Successive Halving as a pruner. The only difference is that we have to set the `pruner` parameter in the `optimize()` method to `optuna.pruners.HyperbandPruner()` in *step 2* in the preceding section. The following code shows you how to perform hyperparameter tuning with the Random Search algorithm as the sampler and HB as the pruner:

```
study = optuna.create_study(direction='maximize',
    sampler=optuna.samplers.RandomSampler(seed=0),
    pruner=optuna.pruners.HyperbandPruner(reduction_factor=3,
min_resource=5)
                              )
```

All of the parameters of HyperbandPruner are the same as SuccessiveHalvingPruner's, except that, here, there is no `min_early_stopping_rate` parameter and there is a `max_resource` parameter. The `min_early_stopping_rate` parameter is removed since it is set automatically based on the ID of each bracket. The `max_resource` parameter is responsible for setting the maximum resource allocated to a trial. By default, this parameter is set to *'auto'*, which means that the value will be set as the largest step in the first completed trial.

Using the exact same data, preprocessing steps, and hyperparameter space, we get around 0.580 of the F1-score evaluated in the validation data. Out of 100 trials performed, there are 79 trials pruned by SH, which implies only 21 completed trials. We also get a dictionary consisting of the best set of hyperparameters as follows:

```
{'num_layers': 0,'optimizer': 'Adam','adam_lr':
0.05584201313189952,'epoch': 37}
```

After the model is trained on full data using the best set of hyperparameters, we get around 0.609 in F1-score when we test the final neural network model trained on the test data.

In this section, we have learned how to utilize HB as a pruner during the hyperparameter tuning process with `Optuna`.

Summary

In this chapter, we have learned all of the important aspects of the `Optuna` package. We have also learned how to implement various hyperparameter tuning methods using the help of this package, in addition to understanding each of the important parameters of the classes and how are they related to the theory that we have learned in previous chapters. From now on, you should be able to utilize the packages we have discussed in the last few chapters to implement your chosen hyperparameter tuning method, and ultimately, boost the performance of your ML model. Equipped with the knowledge from *Chapters 3 - 6*, you will also be able to debug your code if there are errors or unexpected results, and you will be able to craft your own experiment configuration to match your specific problem.

In the next chapter, we will learn about the DEAP and Microsoft NNI packages and how to utilize them to perform various hyperparameter tuning methods. The goal of the next chapter is similar to this chapter, which is to be able to utilize the package for hyperparameter tuning purposes and understand each of the parameters of the implemented classes.

10

Advanced Hyperparameter Tuning with DEAP and Microsoft NNI

DEAP and **Microsoft NNI** are Python packages that provide various hyperparameter tuning methods that are not implemented in other packages that we have discussed in *Chapters 7 – 9*. For example, Genetic Algorithm, Particle Swarm Optimization, Metis, Population-Based Training, and many more.

In this chapter, we'll learn how to perform hyperparameter tuning using both DEAP and Microsoft NNI packages, starting from getting ourselves familiar with the packages, along with the important modules and parameters we need to be aware of. We'll learn not only how to utilize both DEAP and Microsoft NNI to perform hyperparameter tuning with their default configurations but also discuss other available configurations along with their usage. Moreover, we'll also discuss how the implementation of the hyperparameter tuning methods is related to the theory that we have learned in previous chapters, since there may be some minor differences or adjustments made in the implementation.

By the end of this chapter, you will be able to understand all of the important things you need to know about DEAP and Microsoft NNI and be able to implement various hyperparameter tuning methods available in these packages. You'll also be able to understand each of the important parameters of the classes and how they are related to the theory that we have learned in the previous chapters. Finally, equipped with the knowledge from previous chapters, you will also be able to understand what's happening if there are errors or unexpected results and understand how to set up the method configuration to match your specific problem.

The following are the main topics that will be discussed in this chapter:

- Introducing DEAP
- Implementing the Genetic Algorithm
- Implementing Particle Swarm Optimization
- Introducing Microsoft NNI
- Implementing Grid Search
- Implementing Random Search
- Implementing Tree-structured Parzen Estimators
- Implementing Sequential Model Algorithm Configuration
- Implementing Bayesian Optimization Gaussian Process
- Implementing Metis
- Implementing Simulated Annealing
- Implementing Hyper Band
- Implementing Bayesian Optimization Hyper Band
- Implementing Population-Based Training

Technical requirements

We will learn how to implement various hyperparameter tuning methods with DEAP and Microsoft NNI. To ensure that you are able to reproduce the code examples in this chapter, you will require the following:

- Python 3 (version 3.7 or above)
- Installed pandas package (version 1.3.4 or above)
- Installed NumPy package (version 1.21.2 or above)
- Installed SciPy package (version 1.7.3 or above)
- Installed Matplotlib package (version 3.5.0 or above)
- Installed scikit-learn package (version 1.0.1 or above)
- Installed DEAP package (version 1.3)
- Installed Hyperopt package (version 0.1.2)
- Installed NNI package (version 2.7)
- Installed PyTorch package (version 1.10.0)

All of the code examples for this chapter can be found on GitHub at `https://github.com/PacktPublishing/Hyperparameter-Tuning-with-Python/blob/main/10_Advanced_Hyperparameter-Tuning-via-DEAP-and-NNI.ipynb`.

Introducing DEAP

Distributed Evolutionary Algorithms in Python (**DEAP**) is a Python package that allows you to implement various evolutionary algorithms including (but not limited to) the **Genetic Algorithm** (**GA**) and **Particle Swarm Optimization** (**PSO**). To install DEAP, you can simply call the `pip install deap` command.

DEAP allows you to craft your evolutionary algorithm optimization steps in a very flexible manner. The following steps show how to utilize DEAP to perform any hyperparameter tuning methods. More detailed steps, including the code implementation, will be given through various examples in the upcoming sections:

1. Define the *type* classes through the `creator.create()` module. These classes are responsible for defining the type of objects that will be used in the optimization steps.

2. Define the *initializers* along with the hyperparameter space and register them in the `base.Toolbox()` container. The initializers are responsible for setting the initial value of the objects that will be used in the optimization steps.

3. Define the *operators* and register them in the `base.Toolbox()` container. The operators refer to the evolutionary tools or **genetic operator** (see *Chapter 5*) that need to be defined as part of the optimization algorithm. For example, the selection, crossover, and mutation operators in the Genetic Algorithm.

4. Define the objective function and register it in the `base.Toolbox()` container.

5. Define your own hyperparameter tuning algorithm function.

6. Perform hyperparameter tuning by calling the defined function in *step 5*.

7. Train the model on full training data using the best set of hyperparameters found.

8. Test the final trained model on the test data.

The type classes refer to the type of objects used in the optimization steps. These type classes are inherited from the base classes implemented in DEAP. For example, we can define the type of our fitness function as the following:

```
from deap import base, creator
creator.create("FitnessMax", base.Fitness, weights=(1.0,))
```

The base.Fitness class is a base abstract class implemented in DEAP that can be utilized to define our own fitness function type. It expects a weights parameter to understand the type of optimization problem we are working on. If it's a maximization problem, then we have to put a positive weight and the other way around for a minimization problem. Notice that it expects a tuple data structure instead of a float. This is because DEAP also allows us to work with a **multi-objective optimization problem**. So, if we pass (1.0, -1.0) to the weights parameter, it means we have two objective functions where we want to maximize the first one and minimize the second one with equal weight.

The creator.create() function is responsible for creating a new class based on the base class. In the preceding code, we created the type class for our objective function with the name "FitnessMax". This creator.create() function expects at least two parameters: specifically, the name of the newly created class and the base class itself. The rest of the parameters passed to this function will be treated as the attributes for this newly created class. Besides defining the type of the objective function, we can also define the type of individuals in the evolutionary algorithm that will be performed. The following code shows how to create the type of individuals inherited from the built-in list data structure in Python that has fitness as its attribute:

```
creator.create("Individual", list, fitness=creator.FitnessMax)
```

Note that the fitness attribute has a type of creator.FitnessMax, which is the type that we just created in the preceding code.

Types Definition in DEAP

There are a lot of ways to define type classes in DEAP. While we have discussed the most straightforward and, arguably, most used type class, you may find other cases that need other definitions of type class. For more information on how to define other types in DEAP, please refer to the official documentation (https://deap.readthedocs.io/en/master/tutorials/basic/part1.html).

Once we have finished defining the type of objects that will be used in the optimization steps, we now need to initiate the value of those objects using the initializers and register them in the base.Toolbox() container. You can think of this module as a box or container of initializers and other tools that will be utilized during the optimization steps. The following code shows how we can set the random initial values for individuals:

```
import random
from deap import tools
toolbox = base.Toolbox()
toolbox.register("individual", tools.initRepeat, creator.
Individual,
                random.random, n=10)
```

The preceding code shows an example of how to register the `"individual"` object in the `base.Toolbox()` container, where each individual has a size of `10`. The individual is generated by repeatedly calling the `random.random` method 10 times. Note that, in the hyperparameter tuning setup, the size of `10` of each individual actually refers to the number of the hyperparameters we have in the space. The following shows the output of calling the registered individual via the `toolbox.individual()` method:

```
[0.30752039354315985,0.2491982746819209,0.8423374678316783,0.34
01579175109981,0.7699302429041264,0.046433183902334974,0.52870
19598616896,0.28081693679292696,0.9562244184741888,0.000845070
1833065954]
```

As you can see, the output of `toolbox.individual()` is just a list of 10 random values since we've defined `creator.Individual` to inherit from the built-in `list` data structure in Python. Furthermore, we also called `tools.initRepeat` when registering the individual with the `random.random` method by 10 times.

You may now wonder, how do you define the actual hyperparameter space using this `toolbox.register()` method? Initiating a bunch of random values definitely doesn't make any sense. We need to know the way to define the hyperparameter space that will be equipped for each individual. To do that, we can actually utilize another tool provided by DEAP, `tools.InitCycle`.

Where `tools.initRepeat` will just call the provided function n times, in our previous example, the provided function is `random.random`. Here, `tools.InitCycle` expects a list of functions and will call those functions for n cycles. The following code shows an example to define the hyperparameter space that will be equipped for each individual:

1. We need to first register each of the hyperparameters that we have in the space along with their distribution. Note that we can pass all of the required parameters to the sampling distribution function to `toolbox.register()` as well. For example, here, we pass the `a=0,b=0.5,loc=0.005,scale=0.01` parameters of the `truncnorm.rvs()` method:

   ```
   from scipy.stats import randint,truncnorm,uniform
   toolbox.register("param_1", randint.rvs, 5, 200)
   toolbox.register("param_2", truncnorm.rvs, 0, 0.5, 0.005,
   0.01)
   toolbox.register("param_3", uniform.rvs, 0, 1)
   ```

2. Once we have registered each hyperparameter we have, we can register the individual by utilizing `tools.initCycle` with only one cycle of repetition:

   ```
   toolbox.register("individual",tools.initCycle,creator.
   Individual,
       (
   ```

```
            toolbox.param_1,
            toolbox.param_2,
            toolbox.param_3
        ),
        n=1,
    )
```

The following shows the output of calling the registered individual via the `toolbox.individual()` method:

```
[172, 0.005840196235159121, 0.37250162585120816]
```

3. Once we have registered the individual in our toolbox, registering a population is very simple. We just need to utilize the `tools.initRepeat` module and pass the defined `toolbox.individual` as the argument. The following code shows how to register a population in general. Note that, here, the population is just a list of five individuals defined previously:

```
toolbox.register("population", tools.initRepeat, list,
toolbox.individual, n=5)
```

The following shows the output when calling the `toolbox.population()` method:

```
[[168, 0.009384417146554462, 0.4732188841620628],
 [7, 0.009356636359759574, 0.6722125618177741],
 [126, 0.00927973696427319, 0.7417964302134438],
 [88, 0.008112369078803545, 0.4917555243983919],
 [34, 0.008615337472475908, 0.9164442190622125]]
```

As mentioned previously, the `base.Toolbox()` container is responsible for storing not only initializers but also other tools that will be utilized during the optimization steps. Another important building block for an evolutionary algorithm, such as the GA, is the genetic operator. Fortunately, DEAP already implemented various genetic operators that we can utilize via the `tools` module. The following code shows an example of how to register the selection, crossover, and mutation operators for the GA (see *Chapter 5*):

```
# selection strategy
toolbox.register("select", tools.selTournament, tournsize=3)
# crossover strategy
toolbox.register("mate", tools.cxBlend, alpha=0.5)
# mutation strategy
toolbox.register("mutate", tools.mutPolynomialBounded, eta =
0.1, low=-2, up=2, indpb=0.15)
```

The `tools.selTournament` selection strategy works by selecting the best individuals among `tournsize` randomly chosen individuals, *NPOP* times, where `tournsize` is the number of individuals participating in the tournament and *NPOP* is the number of individuals in the population. The `tools.cxBlend` crossover strategy works by performing a linear combination between two continuous individual genes, where the weight for the linear combination is governed by the `alpha` hyperparameter. The `tools.mutPolynomialBounded` mutation strategy works by passing continuous individual genes to a pre-defined polynomial mapping.

Evolutionary Tools in DEAP

There are various built-in evolutionary tools implemented in DEAP that we can utilize for our own needs, starting from initializers, crossover, mutation, selection, and migration tools. For more information regarding the implemented tools, please refer to the official documentation (`https://deap.readthedocs.io/en/master/api/tools.html`).

To register the pre-defined objective function to the toolbox, we can just simply call the same `toolbox.register()` method and pass the objective function, as the following code shows:

```
toolbox.register("evaluate", obj_func)
```

Here, `obj_func` is a Python function that expects to receive the `individual` object defined previously. We will see how to create such an objective function and how to define our own hyperparameter tuning algorithm function in the upcoming sections when we discuss how to implement the GA and PSO in DEAP.

DEAP also allows us to utilize our parallel computing resources when calling the objective function. To do that, we can simply need to register the `multiprocessing` module to the toolbox as the following:

```
import multiprocessing
pool = multiprocessing.Pool()
toolbox.register("map", pool.map)
```

Once we have registered the `multiprocessing` module, we can simply apply this when calling the objective function, as shown in the following code:

```
fitnesses = toolbox.map(toolbox.evaluate, individual)
```

In this section, we have discussed the DEAP package and its building blocks. You may wonder how to construct a real hyperparameter tuning method using all of the building blocks provided by DEAP. Worry no more; in the upcoming two sections, we will learn how to utilize all of the discussed building blocks to perform hyperparameter tuning with the GA and PSO methods.

Implementing the Genetic Algorithm

GA is one of the variants of the Heuristic Search hyperparameter tuning group (see *Chapter 5*) that can be implemented by the DEAP package. To show you how we can implement GA with the DEAP package, let's use the Random Forest classifier model and the same data as in the examples in *Chapter 7*. The dataset used in this example is the *Banking Dataset – Marketing Targets* dataset provided on Kaggle (`https://www.kaggle.com/datasets/prakharrathi25/banking-dataset-marketing-targets`).

The target variable consists of two classes, `yes` or `no`, indicating whether the client of the bank has subscribed to a term deposit or not. Hence, the goal of training an ML model on this dataset is to identify whether a customer is potentially wanting to subscribe to the term deposit or not. Out of the 16 features provided in the data, there are seven numerical features and nine categorical features. As for the target class distribution, 12% of them are *yes* and 88% of them are *no*, for both train and test datasets. For more detailed information about the data, please refer to *Chapter 7*.

Before performing the GA, let's see how the Random Forest classifier with default hyperparameters values works. As shown in *Chapter 7*, we get around `0.436` in the F1-score when evaluating the Random Forest classifier with default hyperparameter values on the test set. Note that we're still using the same scikit-learn pipeline definition to train and evaluate the Random Forest classifier, as explained in *Chapter 7*.

The following code shows how to implement the GA with the DEAP package. You can find the more detailed code in the GitHub repository mentioned in the *Technical requirements* section:

1. Define the GA parameters and type classes through the `creator.create()` module:

    ```
    # GA Parameters
    NPOP = 50 #population size
    NGEN = 15 #number of trials
    CXPB = 0.5 #cross-over probability
    MUTPB = 0.2 #mutation probability
    ```

    Fix the seed for reproducibility:

    ```
    import random
    random.seed(1)
    ```

    Define the type of our fitness function. Here, we are working with a maximization problem and a single objective function, that's why we set `weights=(1.0,)`:

    ```
    from deap import creator, base
    creator.create("FitnessMax", base.Fitness,
    weights=(1.0,))
    ```

Define the type of individuals inherited from the built-in `list` data structure in Python that has `fitness` as its attribute:

```
creator.create("Individual", list, fitness=creator.
FitnessMax)
```

2. Define the initializers along with the hyperparameter space and register them in the `base.Toolbox()` container.

Initialize the toolbox:

```
toolbox = base.Toolbox()
```

Define the naming of the hyperparameters:

```
PARAM_NAMES = ["model__n_estimators","model__criterion",
            "model__class_weight","model__min_samples_
split"
```

Register each of the hyperparameters that we have in the space along with their distribution:

```
from scipy.stats import randint,truncnorm
toolbox.register("model__n_estimators", randint.rvs, 5,
200)
toolbox.register("model__criterion", random.choice,
["gini", "entropy"])
toolbox.register("model__class_weight", random.choice,
["balanced","balanced_subsample"])
toolbox.register("model__min_samples_split", truncnorm.
rvs, 0, 0.5, 0.005, 0.01)
```

Register the individual by utilizing `tools.initCycle` with only one cycle of repetition:

```
from deap import tools
toolbox.register(
    "individual",
    tools.initCycle,
    creator.Individual,
    (
        toolbox.model__n_estimators,
        toolbox.model__criterion,
        toolbox.model__class_weight,
        toolbox.model__min_samples_split,
    ),
)
```

Register the population:

```
toolbox.register("population", tools.initRepeat, list,
toolbox.individual)
```

3.   Define the operators and register them in the base.Toolbox() container.

Register the selection strategy:

```
toolbox.register("select", tools.selTournament,
tournsize=3)
```

Register the cross-over strategy:

```
toolbox.register("mate", tools.cxUniform, indpb=CXPB)
```

Define a custom mutation strategy. Note that all of the implemented mutation strategies in DEAP are not really suitable for hyperparameter tuning purposes since they can only be utilized for floating or binary values, while most of the time, our hyperparameter space will be a combination of real and discrete hyperparameters. The following function shows how to implement such a custom mutation strategy. You can follow the same structure to suit your own need:

```
def mutPolynomialBoundedMix(individual, eta, low, up, is_
int, indpb, discrete_params):
    for i in range(len(individual)):
        if discrete_params[i]:
            if random.random() < indpb:
                individual[i] = random.choice(discrete_
params[i])
        else:
            individual[i] = tools.
mutPolynomialBounded([individual[i]],

eta[i], low[i], up[i], indpb)[0][0]

        if is_int[i]:
            individual[i] = int(individual[i])

    return individual,
```

Register the custom mutation strategy:

```
toolbox.register("mutate", mutPolynomialBoundedMix,
                 eta = [0.1,None,None,0.1],
                 low = [5,None,None,0],
                 up = [200,None,None,1],
                 is_int = [True,False,False,False],
                 indpb=MUTPB,
                 discrete_params=[[],["gini",
"entropy"],["balanced","balanced_subsample"],[]]
                 )
```

4. Define the objective function and register it in the base.Toolbox() container:

```
def evaluate(individual):
    # convert list of parameter values into dictionary of
kwargs
    strategy_params = {k: v for k, v in zip(PARAM_NAMES,
individual)}

    if strategy_params['model__min_samples_split'] > 1 or
strategy_params['model__min_samples_split'] <= 0:
        return [-np.inf]

    tuned_pipe = clone(pipe).set_params(**strategy_
params)
    return [np.mean(cross_val_score(tuned_pipe,X_train_
full, y_train, cv=5, scoring='f1',))]
```

Register the objective function:

```
toolbox.register("evaluate", evaluate)
```

5. Define the Genetic Algorithm with parallel processing:

```
import multiprocessing
import numpy as np
```

Register the multiprocessing module:

```
pool = multiprocessing.Pool(16)
toolbox.register("map", pool.map)
```

Define empty arrays to store the best and average values of objective function scores in each trial:

```
mean = np.ndarray(NGEN)
best = np.ndarray(NGEN)
```

Define a `HallOfFame` class that is responsible for storing the latest best individual (set of hyperparameters) in the population:

```
hall_of_fame = tools.HallOfFame(maxsize=3)
```

Define the initial population:

```
pop = toolbox.population(n=NPOP)
```

Start the GA iterations:

```
for g in range(NGEN):
```

Select the next generation individuals/children/offspring.

```
    offspring = toolbox.select(pop, len(pop))
```

Clone the selected individuals.

```
    offspring = list(map(toolbox.clone, offspring))
```

Apply crossover on the offspring.

```
    for child1, child2 in zip(offspring[::2],
offspring[1::2]):
        if random.random() < CXPB:
            toolbox.mate(child1, child2)
            del child1.fitness.values
            del child2.fitness.values
```

Apply mutation on the offspring.

```
    for mutant in offspring:
        if random.random() < MUTPB:
            toolbox.mutate(mutant)
            del mutant.fitness.values
```

Evaluate the individuals with an invalid fitness.

```
    invalid_ind = [ind for ind in offspring if not ind.
fitness.valid]
    fitnesses = toolbox.map(toolbox.evaluate, invalid_
ind)
    for ind, fit in zip(invalid_ind, fitnesses):
```

```
                        ind.fitness.values = fit
```

The population is entirely replaced by the offspring.

```
        pop[:] = offspring
        hall_of_fame.update(pop)
        fitnesses = [
              ind.fitness.values[0] for ind in pop if not
    np.isinf(ind.fitness.values[0])
        ]
        mean[g] = np.mean(fitnesses)
        best[g] = np.max(fitnesses)
```

6. Perform hyperparameter tuning by running the defined algorithm in *step 5*. After running the GA, we can get the best set of hyperparameters based on the following code:

```
params = {}
for idx_hof, param_name in enumerate(PARAM_NAMES):
    params[param_name] = hall_of_fame[0][idx_hof]
print(params)
```

Based on the preceding code, we get the following results:

```
{'model__n_estimators': 101,
'model__criterion': 'entropy',
'model__class_weight': 'balanced',
'model__min_samples_split': 0.0007106340458649385}
```

We can also plot the trial history or the convergence plot based on the following code:

```
import matplotlib.pyplot as plt
import seaborn as sns
sns.set()
fig, ax = plt.subplots(sharex=True, figsize=(8, 6))
sns.lineplot(x=range(NGEN), y=mean, ax=ax, label="Average
Fitness Score")
sns.lineplot(x=range(NGEN), y=best, ax=ax, label="Best
Fitness Score")
ax.set_title("Fitness Score",size=20)
ax.set_xticks(range(NGEN))
ax.set_xlabel("Iteration")
plt.tight_layout()
plt.show()
```

Based on the preceding code, the following figure is generated. As you can see, the objective function score or the fitness score is increasing throughout the number of trials since the population is updated with the improved individuals:

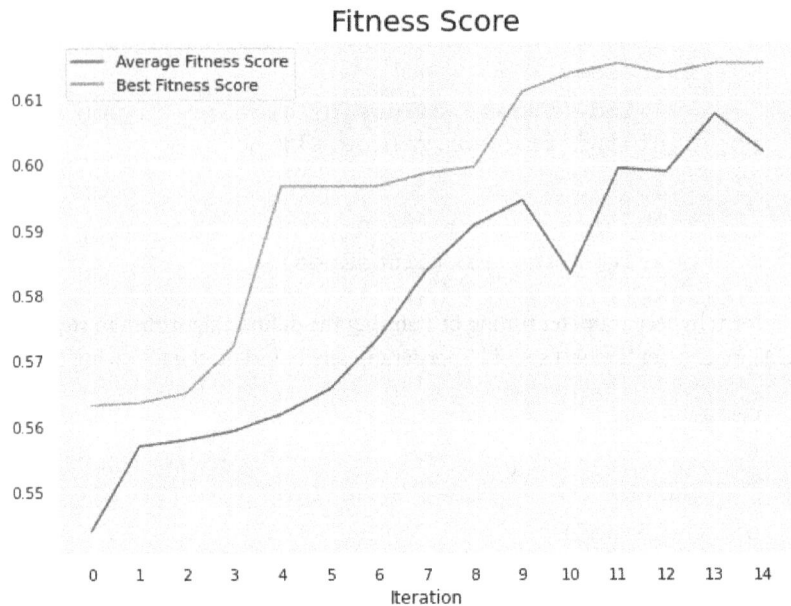

Figure 10.1 – Genetic Algorithm convergence plot

7. Train the model on full training data using the best set of hyperparameters found:

```
from sklearn.base import clone
tuned_pipe = clone(pipe).set_params(**params)
tuned_pipe.fit(X_train_full,y_train)
```

8. Test the final trained model on the test data:

```
y_pred = tuned_pipe.predict(X_test_full)
print(f1_score(y_test, y_pred))
```

Based on the preceding code, we get around 0.608 in the F1-score when testing our final trained Random Forest model with the best set of hyperparameters on the test set.

In this section, we have learned how to implement the GA with the DEAP package, starting from defining the necessary objects and defining the GA procedures with parallel processing and custom mutation strategy, until plotting the history of the trials and testing the best set of hyperparameters in the test set. In the next section, we will learn how to implement the PSO hyperparameter tuning method with the DEAP package.

Implementing Particle Swarm Optimization

PSO is also one of the variants of the Heuristic Search hyperparameter tuning group (see *Chapter 5*) that can be implemented by the DEAP package. We'll still use the same example as in the previous section to see how we can implement PSO using the DEAP package.

The following code shows how to implement PSO with the DEAP package. You can find the more detailed code in the GitHub repository mentioned in the *Technical requirements* section:

1. Define the PSO parameters and type classes through the `creator.create()` module:

    ```
    N = 50 #swarm size
    w = 0.5 #inertia weight coefficient
    c1 = 0.3 #cognitive coefficient
    c2 = 0.5 #social coefficient
    num_trials = 15 #number of trials
    ```

    Fix the seed for reproducibility:

    ```
    import random
    random.seed(1)
    ```

    Define the type of our fitness function. Here, we are working with a maximization problem and a single objective function, which is why we set `weights=(1.0,)`:

    ```
    from deap import creator, base
    creator.create("FitnessMax", base.Fitness,
    weights=(1.0,))
    ```

    Define the type of particles inherited from the built-in `list` data structure in Python that has `fitness`, `speed`, `smin`, `smax`, and `best` as its attribute. These attributes will be utilized later on when updating each particle's position (see *Chapter 5*):

    ```
    creator.create("Particle", list, fitness=creator.
    FitnessMax,
                    speed=list, smin=list, smax=list,
    best=None)
    ```

2. Define the initializers along with the hyperparameter space and register them in the `base.Toolbox()` container.

    Initialize the toolbox:

    ```
    toolbox = base.Toolbox()
    ```

Define the naming of the hyperparameters:

```
PARAM_NAMES = ["model__n_estimators","model__criterion",
               "model__class_weight","model__min_samples_
split"
```

Register each of the hyperparameters that we have in the space along with their distribution. Remember that PSO only works with the numerical type hyperparameters. That's why we encode the "model__criterion" and "model__class_weight" hyperparameters to integers:

```
from scipy.stats import randint,truncnorm
toolbox.register("model__n_estimators", randint.rvs, 5,
200)
toolbox.register("model__criterion", random.choice,
[0,1])
toolbox.register("model__class_weight", random.choice,
[0,1])
toolbox.register("model__min_samples_split", truncnorm.
rvs, 0, 0.5, 0.005, 0.01)
```

Register the individual by utilizing `tools.initCycle` with only one cycle of repetition. Note that we need to also assign the `speed`, `smin`, and `smax` values to each individual. To do that, let's just define a function called `generate`:

```
from deap import tools
def generate(speed_bound):
    part = tools.initCycle(creator.Particle,
                           [toolbox.model__n_estimators,
                            toolbox.model__criterion,
                            toolbox.model__class_weight,
                            toolbox.model__min_samples_
split,
                           ]
                          )
    part.speed = [random.uniform(speed_bound[i]['smin'],
speed_bound[i]['smax']) for i in range(len(part))]
    part.smin = [speed_bound[i]['smin'] for i in
range(len(part))]
    part.smax = [speed_bound[i]['smax'] for i in
range(len(part))]
    return part
```

Register the individual:

```
toolbox.register("particle", generate,
                 speed_bound=[{'smin': -2.5,'smax': 2.5},
                             {'smin': -1,'smax': 1},
                             {'smin': -1,'smax': 1},
                             {'smin': -0.001,'smax':
0.001}])
```

Register the population:

```
toolbox.register("population", tools.initRepeat, list,
toolbox.particle)
```

3. Define the operators and register them in the base.Toolbox() container. The main operator in PSO is the particle's position update operator, which is defined in the updateParticle function as follows:

```
import operator
import math
def updateParticle(part, best, c1, c2, w, is_int):
    w = [w for _ in range(len(part))]
    u1 = (random.uniform(0, 1)*c1 for _ in
range(len(part)))
    u2 = (random.uniform(0, 1)*c2 for _ in
range(len(part)))
    v_u1 = map(operator.mul, u1, map(operator.sub, part.
best, part))
    v_u2 = map(operator.mul, u2, map(operator.sub, best,
part))
    part.speed = list(map(operator.add, map(operator.mul,
w, part.speed), map(operator.add, v_u1, v_u2)))
    for i, speed in enumerate(part.speed):
        if abs(speed) < part.smin[i]:
            part.speed[i] = math.copysign(part.smin[i],
speed)
        elif abs(speed) > part.smax[i]:
            part.speed[i] = math.copysign(part.smax[i],
speed)
    part[:] = list(map(operator.add, part, part.speed))
```

```
for i, pos in enumerate(part):
    if is_int[i]:
        part[i] = int(pos)
```

Register the operator. Note that the is_int attribute is responsible for marking which hyperparameter has an integer type of value:

```
toolbox.register("update", updateParticle, c1=c1, c2=c2,
w=w,
                is_int=[True,True,True,False]
                )
```

4. Define the objective function and register it in the base.Toolbox() container. Note that we also decode the "model__criterion" and "model__class_weight" hyperparameters within the objective function:

```
def evaluate(particle):
    # convert list of parameter values into dictionary of
kwargs
    strategy_params = {k: v for k, v in zip(PARAM_NAMES,
particle)}
    strategy_params["model__criterion"] = "gini" if
strategy_params["model__criterion"]==0 else "entropy"
    strategy_params["model__class_weight"] = "balanced"
if strategy_params["model__class_weight"]==0 else
"balanced_subsample"

    if strategy_params['model__min_samples_split'] > 1 or
strategy_params['model__min_samples_split'] <= 0:
        return [-np.inf]

    tuned_pipe = clone(pipe).set_params(**strategy_
params)

    return [np.mean(cross_val_score(tuned_pipe,X_train_
full, y_train, cv=5, scoring='f1',))]
```

Register the objective function:

```
toolbox.register("evaluate", evaluate)
```

5. Define PSO with parallel processing:

```
import multiprocessing
import numpy as np
```

Register the multiprocessing module:

```
pool = multiprocessing.Pool(16)
toolbox.register("map", pool.map)
```

Define empty arrays to store the best and average values of objective function scores in each trial:

```
mean_arr = np.ndarray(num_trials)
best_arr = np.ndarray(num_trials)
```

Define a HallOfFame class that is responsible for storing the latest best individual (set of hyperparameters) in the population:

```
hall_of_fame = tools.HallOfFame(maxsize=3)
```

Define the initial population:

```
pop = toolbox.population(n=NPOP)
```

Start the PSO iterations:

```
best = None
for g in range(num_trials):
    fitnesses = toolbox.map(toolbox.evaluate, pop)
    for part, fit in zip(pop, fitnesses):
        part.fitness.values = fit


        if not part.best or part.fitness.values > part.
best.fitness.values:
            part.best = creator.Particle(part)
            part.best.fitness.values = part.fitness.
values
        if not best or part.fitness.values > best.
fitness.values:
            best = creator.Particle(part)
            best.fitness.values = part.fitness.values
    for part in pop:
        toolbox.update(part, best)
```

```
        hall_of_fame.update(pop)
        fitnesses = [
                ind.fitness.values[0] for ind in pop if not
    np.isinf(ind.fitness.values[0])
        ]
        mean_arr[g] = np.mean(fitnesses)
        best_arr[g] = np.max(fitnesses)
```

6. Perform hyperparameter tuning by running the algorithm defined in *step 5*. After running PSO, we can get the best set of hyperparameters based on the following code. Note that we need to decode the "model__criterion" and "model__class_weight" hyperparameters before passing them to the final model:

```
params = {}
for idx_hof, param_name in enumerate(PARAM_NAMES):
    if param_name == "model__criterion":
        params[param_name] = "gini" if hall_of_fame[0]
    [idx_hof]==0 else "entropy"
    elif param_name == "model__class_weight":
        params[param_name] = "balanced" if hall_of_
    fame[0][idx_hof]==0 else "balanced_subsample"
    else:
        params[param_name] = hall_of_fame[0][idx_hof]
print(params)
```

Based on the preceding code, we get the following results:

```
{'model__n_estimators': 75,
 'model__criterion': 'entropy',
 'model__class_weight': 'balanced',
 'model__min_samples_split': 0.0037241038302412493}
```

7. Train the model on full training data using the best set of hyperparameters found:

```
from sklearn.base import clone
tuned_pipe = clone(pipe).set_params(**params)
tuned_pipe.fit(X_train_full,y_train)
```

8. Test the final trained model on the test data:

```
y_pred = tuned_pipe.predict(X_test_full)
print(f1_score(y_test, y_pred))
```

Based on the preceding code, we get around 0.569 in the F1-score when testing our final trained Random Forest model with the best set of hyperparameters on the test set.

In this section, we have learned how to implement PSO with the DEAP package, starting from defining the necessary objects, encoding the categorical hyperparameter to integers, and defining the optimization procedures with parallel processing, until testing the best set of hyperparameters in the test set. In the next section, we will start learning about another hyperparameter tuning package called NNI, which is developed by Microsoft.

Introducing Microsoft NNI

Neural Network Intelligence (**NNI**) is a package that is developed by Microsoft and can be utilized not only for hyperparameter tuning tasks but also for neural architecture search, model compression, and feature engineering. In this section, we will discuss how to utilize NNI specifically for the hyperparameter tuning task. To install NNI, you can simply call the `pip install nni` command.

Although NNI refers to *Neural Network Intelligence*, it actually supports numerous ML frameworks including (but not limited to) scikit-learn, XGBoost, LightGBM, PyTorch, TensorFlow, Caffe2, and MXNet.

There are numerous hyperparameter tuning methods implemented in NNI; some of them are built-in and others are wrapped from other packages such as `Hyperopt` (see *Chapter 8*) and `SMAC3`. Here, in NNI, the hyperparameter tuning methods are referred to as **tuners**. We will not discuss all of the tuners implemented in NNI since there are too many of them. We will only discuss the tuners that have been discussed in *Chapters 3 – 6*. Apart from tuners, some of the hyperparameter tuning methods, such as Hyper Band and BOHB, are treated as **advisors** in NNI.

Available Tuners in NNI

To see all of the available tuners in NNI, please refer to the official documentation page (`https://nni.readthedocs.io/en/stable/hpo/tuners.html`).

Unlike other hyperparameter tuning packages that we have discussed so far, in NNI, we have to prepare a Python script containing the model definition before being able to run the hyperparameter tuning process from the notebook. Furthermore, NNI also allows us to run the hyperparameter tuning experiment from the command-line tool where we need to define several other additional files to store the hyperparameter space information and other configurations.

The following steps show how we can perform any hyperparameter tuning procedure with NNI with pure Python code:

1. Prepare the model to be tuned in a script, for example, `model.py`. This script should include the model architecture definition, dataset loading function, training function, and testing function. It also has to include three NNI API calls, as follows:

    *   `nni.get_next_parameter()` is responsible for collecting the hyperparameters to be evaluated in a particular trial.

    *   `nni.report_intermediate_result()` is responsible for reporting the evaluation metric within each training iteration (epoch or steps). Note that this API call is not mandatory; if you can't get the intermediate evaluation metric from your ML framework, then this API call is not required.

    *   `nni.report_final_result()` is responsible for reporting the final evaluation metric score after the training process is finished.

2. Define the hyperparameter space. NNI expects the hyperparameter space is in the form of a Python dictionary, where the first-level keys store the names of the hyperparameters. The second-level keys store the types of the sampling distribution and the hyperparameter values range. The following shows an example of how to define the hyperparameter space in the expected format:

    ```
    hyperparameter_space = {
        'n_estimators': {'_type': 'randint', '_value': [5,
    200]},
        'criterion': {'_type': 'choice', '_value': ['gini',
    'entropy']},
        'min_samples_split': {'_type': 'uniform', '_value':
    [0, 0.1]},
    }
    ```

> **More Information on NNI**
>
> For more information regarding the supported sampling distributions in NNI, please refer to the official documentation (https://nni.readthedocs.io/en/latest/hpo/search_space.html).

3. Next, we need to set up the experiment configurations via the `Experiment` class. The following shows steps to set up several configurations before we can run the hyperparameter tuning process.

Load the `Experiment` class. Here, we are using the `'local'` experiment mode, which means all the training and hyperparameter tuning processes will be done only on our local computer. NNI allows us to run the training procedures in various platforms, including (but not limited to) **Azure Machine Learning** (**AML**), Kubeflow, and OpenAPI. For more information, please refer to the official documentation (`https://nni.readthedocs.io/en/latest/reference/experiment_config.html`):

```
from nni.experiment import Experiment
experiment = Experiment('local')
```

Set up the trial code configuration. Here, we need to specify the command to run the defined script in *step 1* and the relative path to the script. The following shows an example of how to set up the trial code configuration:

```
experiment.config.trial_command = 'python model.py'
experiment.config.trial_code_directory = '.'
```

Set up the hyperparameter space configuration. To set up the hyperparameter space configuration, we simply need to pass the defined hyperparameter space in *step 2*. The following code shows how to do that:

```
experiment.config.search_space = hyperparameter_space
```

Set up the hyperparameter tuning algorithm to be utilized. The following shows an example of how to use TPE as the hyperparameter tuning algorithm on a maximization problem:

```
experiment.config.tuner.name = 'TPE'
experiment.config.tuner.class_args['optimize_mode'] = 'maximize'
```

Set up the number of trials and concurrent processes. NNI allows us to set how many numbers of hyperparameter sets are to be evaluated concurrently at a single time. The following code shows an example of how to set the number of trials to 50, where five sets will be evaluated concurrently at a particular time:

```
experiment.config.max_trial_number = 50
experiment.config.trial_concurrency = 5
```

It is worth noting that NNI also allows you to define the stopping criterion based on the time duration instead of the number of trials. The following code shows how you can set the limit of the experiment time to 1 hour:

```
experiment.config.max_experiment_duration = '1h'
```

If you don't provide both `max_trial_number` and `max_experiment_duration`, then the experiment will run forever until you forcefully stop it via the *Ctrl + C* command.

4. Run the hyperparameter tuning experiment. To run the experiment, we can simply call the run method on the Experiment class. Here, we have to also choose what port to be used. We can see the experiment status and various interesting stats via the launched web portal. The following code shows how to run the experiment on port 8080 in local, meaning you can open the web portal on http://localhost:8080:

    ```
    experiment.run(8080)
    ```

    There are two available Boolean parameters for the run method, namely wait_completion and debug. When we set wait_completion=True, we can't run other cells in the notebook until the experiment is done or some errors are found. The debug parameter enables us to choose whether we want to start the experiment in debug mode or not.

5. Train the model on full training data using the best set of hyperparameters found.

6. Test the final trained model on the test data.

NNI Web Portal

For more information regarding features available in the web portal, please refer to the official documentation (https://nni.readthedocs.io/en/stable/experiment/web_portal/web_portal.html). Note that we will discuss the web portal more in *Chapter 13, Tracking Hyperparameter Tuning Experiments*.

If you prefer to work with the command-line tool, the following steps show how to perform any hyperparameter tuning procedure with NNI with the command-line tool, JSON, and YAML config files:

1. Prepare the model to be tuned in a script. This step is exactly the same as the previous procedure to perform hyperparameter tuning with NNI with pure Python code.

2. Define the hyperparameter space. The expected format of the hyperparameter space is exactly the same as in the procedure on how to perform any hyperparameter tuning procedure with NNI with pure Python code. However, here, we need to store the Python dictionary within a JSON file, for example, hyperparameter_space.json.

3. Set up the experiment configurations via the config.yaml file. The configurations that need to be set up are basically the same as in the procedure with NNI with pure Python code. However, instead of configuring the experiment via a Python class, here, we store all of the configuration details in a single YAML file. The following shows an example of what the YAML file will look like:

```
searchSpaceFile: hyperparameter_space.json

trial_command: python model.py
trial_code_directory: .

trial_concurrency: 5
max_trial_number: 50

tuner:
  name: TPE
  class_args:
    optimize_mode: maximize

training_service:
  platform: local
```

4. Run the hyperparameter tuning experiment. To run the experiment, we can simply call the `nnictl create` command. The following code shows how to use the command to run the experiment on port `8080` in `local`:

```
nnictl create --config config.yaml --port 8080
```

When the experiment is done, you can easily stop the process via the `nnictl stop` command.

5. Train the model on full training data using the best set of hyperparameters found.

6. Test the final trained model on the test data.

Examples for Various ML Frameworks

You can find all of the examples to perform hyperparameter tuning via NNI using your favorite ML frameworks in the official documentation (`https://github.com/microsoft/nni/tree/master/examples/trials`).

---

scikit-nni

There is also a package called `scikit-nni`, which will automatically generate the required `config.yml` and `search-space.json` and build the `scikit-learn` pipelines based on your own custom needs. Please refer to the official repository for further information about this package (`https://github.com/ksachdeva/scikit-nni`).

Besides tuners or hyperparameter tuning algorithms, NNI also provides **assessors** that can be utilized. Assessors are basically early stopping modules that can be used to control the hyperparameter tuning experiment when there's a sign that we may not need to finish the whole experiment trials. Assessors can only be utilized when we provide the intermediate results to NNI via the `nni.report_intermediate_result()` API call. There are only two built-in assessors in NNI: *median stop* and *curve fitting*. The first assessor will stop the experiment whenever a hyperparameter set performs worse than the median at any step. The latter assessor will stop the experiment if the learning curve is likely to converge to a suboptimal result.

Setting up an assessor in NNI is very straightforward. You can simply add the configuration on the `Experiment` class or within the `config.yaml` file. The following code shows how to configure the median stop assessor on the `Experiment` class:

```
experiment.config.assessor.name = 'Medianstop'
```

> **Custom Algorithms in NNI**
>
> NNI also allows us to define our own custom tuners and assessors. To do that, you need to inherit the base `Tuner` or `Assessor` class, write several required functions, and add more details on the `Experiment` class or in the `config.yaml` file. For more information regarding how to define your own custom tuners and assessors, please refer to the official documentation (`https://nni.readthedocs.io/en/stable/hpo/custom_algorithm.html`).

In this section, we have discussed the NNI package and how to perform hyperparameter tuning experiments in general. In the upcoming sections, we will learn how to implement various hyperparameter tuning algorithms using NNI.

Implementing Grid Search

Grid Search is one of the variants of the Exhaustive Search hyperparameter tuning group (see *Chapter 3*) that the NNI package can implement. To show you how we can implement Grid Search with the NNI package, let's use the same data and pipeline as in the examples in the previous section. However, here, we'll define a new hyperparameter space since NNI supports only limited types of sampling distribution.

The following code shows how to implement Grid Search with the NNI package. Here, we'll use the NNI command-line tool (**nnictl**) instead of using pure Python code. You can find the more detailed code in the GitHub repository mentioned in the *Technical requirements* section:

1. Prepare the model to be tuned in a script. Here, we'll name the script `model.py`. There are several functions defined within this script, including `load_data`, `get_default_parameters`, `get_model`, and `run`.

The `load_data` function loads the original data and splits it into train and test data. Furthermore, it's also responsible for returning the lists of numerical and categorical column names:

```
import pandas as pd
import numpy as np
from sklearn.model_selection import train_test_split
from pathlib import Path
def load_data():
    df = pd.read_csv(f"{Path(__file__).parent.parent}/
train.csv",sep=";")

    #Convert the target variable to integer
    df['y'] = df['y'].map({'yes':1,'no':0})

    #Split full data into train and test data
    df_train, df_test = train_test_split(df, test_
size=0.1, random_state=0)

    #Get list of categorical and numerical features
    numerical_feats = list(df_train.drop(columns='y').
select_dtypes(include=np.number).columns)
    categorical_feats = list(df_train.drop(columns='y').
select_dtypes(exclude=np.number).columns)

    X_train - df_train.drop(columns=['y'])
    y_train = df_train['y']
    X_test = df_test.drop(columns=['y'])
    y_test = df_test['y']

    return X_train, X_test, y_train, y_test, numerical_
feats, categorical_feats
```

The `get_default_parameters` function returns the default hyperparameter values used in the experiment:

```
def get_default_parameters():
    params = {
        'model__n_estimators': 5,
        'model__criterion': 'gini',
```

```
        'model__class_weight': 'balanced',
        'model__min_samples_split': 0.01,
    }

    return params
```

The get_model function defines the sklearn pipeline used in this example:

```
from sklearn.compose import ColumnTransformer
from sklearn.preprocessing import StandardScaler,
OneHotEncoder
from sklearn.pipeline import Pipeline
from sklearn.ensemble import RandomForestClassifier
def get_model(PARAMS, numerical_feats, categorical_
feats):
```

Initiate the Normalization Pre-processing for Numerical Features.

```
    numeric_preprocessor = StandardScaler()
```

Initiate the One-Hot-Encoding Pre-processing for Categorical Features.

```
    categorical_preprocessor = OneHotEncoder(handle_
unknown="ignore")
```

Create the ColumnTransformer Class to delegate each preprocessor to the corresponding features.

```
    preprocessor = ColumnTransformer(
        transformers=[
            ("num", numeric_preprocessor, numerical_
feats),
            ("cat", categorical_preprocessor,
categorical_feats),
        ]
    )
```

Create a Pipeline of preprocessor and model.

```
    pipe = Pipeline(
        steps=[("preprocessor", preprocessor),
                ("model", RandomForestClassifier(random_
state=0))]
    )
```

Set hyperparmeter values.

```
pipe = pipe.set_params(**PARAMS)


return pipe
```

The run function is responsible for training the model and getting the cross-validation score:

```
import nni
import logging
from sklearn.model_selection import cross_val_score
LOG = logging.getLogger('nni_sklearn')
def run(X_train, y_train, model):
    model.fit(X_train, y_train)
    score = np.mean(cross_val_score(model,X_train, y_
train,
                    cv=5, scoring='f1')
            )
    LOG.debug('score: %s', score)
    nni.report_final_result(score)
```

Finally, we can call those functions in the same script:

```
if __name__ == '__main__':
    X_train, _, y_train, _, numerical_feats, categorical_
feats = load_data()
    try:
        # get parameters from tuner
        RECEIVED_PARAMS = nni.get_next_parameter()
        LOG.debug(RECEIVED_PARAMS)
        PARAMS = get_default_parameters()
        PARAMS.update(RECEIVED_PARAMS)
        LOG.debug(PARAMS)
        model = get_model(PARAMS, numerical_feats,
categorical_feats)
        run(X_train, y_train, model)
    except Exception as exception:
        LOG.exception(exception)
        raise
```

2. Define the hyperparameter space in a JSON file called `hyperparameter_space.json`:

```
{"model__n_estimators": {"_type": "randint", "_value":
[5, 200]}, "model__criterion": {"_type": "choice", "_
value": ["gini", "entropy"]}, "model__class_weight":
{"_type": "choice", "_value": ["balanced","balanced_
subsample"]}, "model__min_samples_split": {"_type":
"uniform", "_value": [0, 0.1]}}
```

3. Set up the experiment configurations via the `config.yaml` file:

```
searchSpaceFile: hyperparameter_space.json

experimentName: nni_sklearn

trial_command: python '/mnt/c/Users/Louis\ Owen/Desktop/
Packt/Hyperparameter-Tuning-with-Python/nni/model.py'

trial_code_directory:  .

trial_concurrency: 10

max_trial_number: 100

maxExperimentDuration: 1h

tuner:
    name: GridSearch

training_service:
    platform: local
```

4. Run the hyperparameter tuning experiment. We can see the experiment status and various interesting stats via the launched web portal. The following code shows how to run the experiment on port `8080` in `local`, meaning you can open the web portal on `http://localhost:8080`:

```
nnictl create --config config.yaml --port 8080
```

5. Train the model on full training data using the best set of hyperparameters found. To get the best set of hyperparameters, you can go to the web portal and see them from the **Overview** tab.

   Based on the experiment results shown in the web portal within the *Top trials* tab, the following are the best hyperparameter values found from the experiment. Note that we will discuss the web portal more in *Chapter 13*, *Tracking Hyperparameter Tuning Experiments*:

```
best_parameters = {
    "model__n_estimators": 27,
    "model__criterion": "entropy",
    "model__class_weight": "balanced_subsample",
```

```
      "model__min_samples_split": 0.05
    }
```

We can now train the model on full training data:

```
from sklearn.base import clone
tuned_pipe = clone(pipe).set_params(**best_parameters)
# Fit the pipeline on train data
tuned_pipe.fit(X_train_full,y_train)
```

6. Test the final trained model on the test data:

```
y_pred = tuned_pipe.predict(X_test_full)
print(f1_score(y_test, y_pred))
```

Based on the preceding code, we get around 0.517 in the F1-score when testing our final trained Random Forest model with the best set of hyperparameters on the test set.

In this section, we have learned how to implement Grid Search with the NNI package via `nnictl`. In the next section, we will learn how to implement Random Search with NNI via pure Python code.

Implementing Random Search

Random Search is one of the variants of the Exhaustive Search hyperparameter tuning group (see *Chapter 3*) that the NNI package can implement. Let's use the same data, pipeline, and hyperparameter space as in the example in the previous section to show you how to implement Random Search with NNI using pure Python code.

The following code shows how to implement Random Search with the NNI package. Here, we'll use pure Python code instead of using `nnictl` as in the previous section. You can find the more detailed code in the GitHub repository mentioned in the *Technical requirements* section:

1. Prepare the model to be tuned in a script. We'll use the same `model.py` script as in the previous section.

2. Define the hyperparameter space in the form of a Python dictionary:

```
hyperparameter_space = {
    'model__n_estimators': {'_type': 'randint', '_value':
[5, 200]},
    'model__criterion': {'_type': 'choice', '_value':
['gini', 'entropy']},
    'model__class_weight': {'_type': 'choice', '_value':
["balanced","balanced_subsample"]},
    'model__min_samples_split': {'_type': 'uniform', '_
```

```
value': [0, 0.1]},
}
```

3. Set up the experiment configurations via the `Experiment` class. Note that there is only one parameter for the Random Search tuner, namely the random `seed` parameter:

    ```
    experiment = Experiment('local')


    experiment.config.experiment_name = 'nni_sklearn_random_
    search'
    experiment.config.tuner.name = 'Random'
    experiment.config.tuner.class_args['seed'] = 0


    # Boilerplate code
    experiment.config.trial_command = "python '/mnt/c/Users/
    Louis\ Owen/Desktop/Packt/Hyperparameter-Tuning-with-
    Python/nni/model.py'"
    experiment.config.trial_code_directory = '.'
    experiment.config.search_space = hyperparameter_space
    experiment.config.max_trial_number = 100
    experiment.config.trial_concurrency = 10
    experiment.config.max_experiment_duration = '1h'
    ```

4. Run the hyperparameter tuning experiment:

    ```
    experiment.run(8080, wait_completion = True, debug =
    False)
    ```

5. Train the model on full training data using the best set of hyperparameters found.

    Get the best set of hyperparameters:

    ```
    best_trial = sorted(experiment.export_data(),key = lambda
    x: x.value, reverse = True)[0]
    print(best_trial.parameter)
    ```

6. Based on the preceding code, we get the following results:

    ```
    {'model__n_estimators': 194, 'model__criterion':
    'entropy', 'model__class_weight': 'balanced_subsample',
    'model__min_samples_split': 0.0014706304965369289}
    ```

We can now train the model on full training data:

```
from sklearn.base import clone
tuned_pipe = clone(pipe).set_params(**best_trial.
parameter)
# Fit the pipeline on train data
tuned_pipe.fit(X_train_full,y_train)
```

7. Test the final trained model on the test data:

```
y_pred = tuned_pipe.predict(X_test_full)
print(f1_score(y_test, y_pred))
```

Based on the preceding code, we get around `0.597` in the F1-score when testing our final trained Random Forest model with the best set of hyperparameters on the test set.

In this section, we have learned how to implement Random Search using NNI with pure Python code. In the next section, we will learn how to implement Tree-structured Parzen Estimators with NNI via pure Python code.

Implementing Tree-structured Parzen Estimators

Tree-structured Parzen Estimators (**TPEs**) are one of the variants of the Bayesian Optimization hyperparameter tuning group (see *Chapter 4*) that the NNI package can implement. Let's use the same data, pipeline, and hyperparameter space as in the example in the previous section to implement TPE with NNI using pure Python code.

The following code shows how to implement TPE with the NNI package using pure Python code. You can find the more detailed code in the GitHub repository mentioned in the *Technical requirements* section:

1. Prepare the model to be tuned in a script. We'll use the same `model.py` script as in the previous section.

2. Define the hyperparameter space in the form of a Python dictionary. We'll use the same hyperparameter space as in the previous section.

3. Set up the experiment configurations via the `Experiment` class. Note that there are three parameters for the TPE tuner: `optimize_mode`, `seed`, and `tpe_args`. Please refer to the official documentation page for more information regarding the TPE tuner parameters (`https://nni.readthedocs.io/en/stable/reference/hpo.html#tpe-tuner`):

```
experiment = Experiment('local')

experiment.config.experiment_name = 'nni_sklearn_tpe'
experiment.config.tuner.name = 'TPE'
```

```
experiment.config.tuner.class_args = {'optimize_mode':
'maximize', 'seed': 0}


# Boilerplate code
# same with previous section
```

4. Run the hyperparameter tuning experiment:

```
experiment.run(8080, wait_completion = True, debug =
False)
```

5. Train the model on full training data using the best set of hyperparameters found.

Get the best set of hyperparameters:

```
best_trial = sorted(experiment.export_data(),key = lambda
x: x.value, reverse = True)[0]
print(best_trial.parameter)
```

Based on the preceding code, we get the following results:

```
{'model__n_estimators': 195, 'model__criterion':
'entropy', 'model__class_weight': 'balanced_subsample',
'model__min_samples_split': 0.0006636374717157983}
```

We can now train the model on full training data:

```
from sklearn.base import clone
tuned_pipe = clone(pipe).set_params(**best_trial.
parameter)
```

Fit the pipeline on train data.

```
tuned_pipe.fit(X_train_full,y_train)
```

6. Test the final trained model on the test data:

```
y_pred = tuned_pipe.predict(X_test_full)
print(f1_score(y_test, y_pred))
```

Based on the preceding code, we get around 0.618 in the F1-score when testing our final trained Random Forest model with the best set of hyperparameters on the test set.

In this section, we have learned how to implement TPE using NNI with pure Python code. In the next section, we will learn how to implement Sequential Model Algorithm Configuration with NNI via pure Python code.

Implementing Sequential Model Algorithm Configuration

Sequential Model Algorithm Configuration (**SMAC**) is one of the variants of the Bayesian Optimization hyperparameter tuning group (see *Chapter 4*) that the NNI package can implement. Note that to use SMAC in NNI, we need to install additional dependencies using the following command: `pip install "nni[SMAC]"`. Let's use the same data, pipeline, and hyperparameter space as in the example in the previous section to implement SMAC with NNI using pure Python code.

The following code shows how to implement SMAC with the NNI package using pure Python code. You can find the more detailed code in the GitHub repository mentioned in the *Technical requirements* section:

1. Prepare the model to be tuned in a script. We'll use the same `model.py` script as in the previous section.

2. Define the hyperparameter space in the form of a Python dictionary. We'll use the same hyperparameter space as in the previous section.

3. Set up the experiment configurations via the `Experiment` class. Note that there are two parameters for the SMAC tuner: `optimize_mode`, and `config_dedup`. Please refer to the official documentation page for more information regarding the SMAC tuner parameters (`https://nni.readthedocs.io/en/stable/reference/hpo.html#smac-tuner`):

   ```
   experiment = Experiment('local')

   experiment.config.experiment_name = 'nni_sklearn_smac'
   experiment.config.tuner.name = 'SMAC'
   experiment.config.tuner.class_args['optimize_mode'] =
   'maximize'
   # Boilerplate code
   # same with previous section
   ```

4. Run the hyperparameter tuning experiment:

   ```
   experiment.run(8080, wait_completion = True, debug =
   False)
   ```

5. Train the model on full training data using the best set of hyperparameters found.

   Get the best set of hyperparameters:
   ```
   best_trial = sorted(experiment.export_data(),key = lambda
   x: x.value, reverse = True)[0]
   print(best_trial.parameter)
   ```

Based on the preceding code, we get the following results:

```
{'model__class_weight': 'balanced', 'model__
criterion': 'entropy', 'model__min_samples_split':
0.0005502416428725066, 'model__n_estimators': 199}
```

We can now train the model on full training data:

```
from sklearn.base import clone
tuned_pipe = clone(pipe).set_params(**best_trial.
parameter)
# Fit the pipeline on train data
tuned_pipe.fit(X_train_full,y_train)
```

6. Test the final trained model on the test data:

```
y_pred = tuned_pipe.predict(X_test_full)
print(f1_score(y_test, y_pred))
```

Based on the preceding code, we get around 0.619 in the F1-score when testing our final trained Random Forest model with the best set of hyperparameters on the test set.

In this section, we have learned how to implement SMAC using NNI with pure Python code. In the next section, we will learn how to implement Bayesian Optimization Gaussian Process with NNI via pure Python code.

Implementing Bayesian Optimization Gaussian Process

Bayesian Optimization Gaussian Process (**BOGP**) is one of the variants of the Bayesian Optimization hyperparameter tuning group (see *Chapter 4*) that the NNI package can implement. Let's use the same data, pipeline, and hyperparameter space as in the example in the previous section to implement BOGP with NNI using pure Python code.

The following code shows how to implement BOGP with the NNI package using pure Python code. You can find the more detailed code in the GitHub repository mentioned in the *Technical requirements* section:

1. Prepare the model to be tuned in a script. Here, we'll use a new script called model_numeric. py. In this script, we add a mapping for non-numeric hyperparameters since BOGP can only work with numerical hyperparameters:

```
non_numeric_mapping = params = {
    'model__criterion': ['gini','entropy'],
    'model__class_weight': ['balanced','balanced_
subsample'],
    }
```

2. Define the hyperparameter space in the form of a Python dictionary. We'll use a similar hyperparameter space as in the previous section with the only difference on the non-numeric hyperparameters. Here, all of the non-numeric hyperparameters are encoded into integer types of values:

```
hyperparameter_space_numeric = {
    'model__n_estimators': {'_type': 'randint', '_value':
[5, 200]},
    'model__criterion': {'_type': 'choice', '_value': [0,
1]},
    'model__class_weight': {'_type': 'choice', '_value':
[0, 1]},
    'model__min_samples_split': {'_type': 'uniform', '_
value': [0, 0.1]},
}
```

3. Set up the experiment configurations via the Experiment class. Note that there are nine parameters for the BOGP tuner: optimize_mode, utility, kappa, xi, nu, alpha, cold_start_num, selection_num_warm_up, and selection_num_starting_points. Please refer to the official documentation page for more information regarding the BOGP tuner parameters (https://nni.readthedocs.io/en/stable/reference/hpo.html#gp-tuner):

```
experiment = Experiment('local')

experiment.config.experiment_name = 'nni_sklearn_bogp'
experiment.config.tuner.name = 'GPTuner'
experiment.config.tuner.class_args = {
'optimize_mode': 'maximize', 'utility': 'ei','xi': 0.01}

# Boilerplate code
experiment.config.trial_command = "python '/mnt/c/Users/
Louis\ Owen/Desktop/Packt/Hyperparameter-Tuning-with-
Python/nni/model_numeric.py'"
experiment.config.trial_code_directory = '.'
experiment.config.search_space = hyperparameter_space_
numeric
experiment.config.max_trial_number = 100
experiment.config.trial_concurrency = 10
experiment.config.max_experiment_duration = '1h'
```

4. Run the hyperparameter tuning experiment:

    ```
    experiment.run(8080, wait_completion = True, debug =
    False)
    ```

5. Train the model on full training data using the best set of hyperparameters found.

    Get the best set of hyperparameters:

    ```
    non_numeric_mapping = params = {
    'model__criterion': ['gini','entropy'],
    'model__class_weight': ['balanced','balanced_subsample'],
        }
    best_trial = sorted(experiment.export_data(),key = lambda
    x: x.value, reverse = True)[0]
    for key in non_numeric_mapping:
        best_trial.parameter[key] = non_numeric_mapping[key]
    [best_trial.parameter[key]]
    print(best_trial.parameter)
    ```

    Based on the preceding code, we get the following results:

    ```
    {'model__class_weight': 'balanced_subsample', 'model__
    criterion': 'entropy', 'model__min_samples_split':
    0.00055461211818435, 'model__n_estimators': 159}
    ```

    We can now train the model on full training data:

    ```
    from sklearn.base import clone
    tuned_pipe = clone(pipe).set_params(**best_trial.
    parameter)
    ```

    Fit the pipeline on train data.

    ```
    tuned_pipe.fit(X_train_full,y_train)
    ```

6. Test the final trained model on the test data:

    ```
    y_pred = tuned_pipe.predict(X_test_full)
    print(f1_score(y_test, y_pred))
    ```

    Based on the preceding code, we get around 0.619 in the F1-score when testing our final trained Random Forest model with the best set of hyperparameters on the test set.

In this section, we have learned how to implement BOGP using NNI with pure Python code. In the next section, we will learn how to implement Metis with NNI via pure Python code.

Implementing Metis

Metis is one of the variants of the Bayesian Optimization hyperparameter tuning group (see *Chapter 4*) that the NNI package can implement. Let's use the same data, pipeline, and hyperparameter space as in the example in the previous section to implement Metis with NNI using pure Python code.

The following code shows how to implement Metis with the NNI package using pure Python code. You can find the more detailed code in the GitHub repository mentioned in the *Technical requirements* section:

1. Prepare the model to be tuned in a script. Here, we'll use the same script as in the previous section, `model_numeric.py`, since Metis can only work with numerical hyperparameters.

2. Define the hyperparameter space in the form of a Python dictionary. We'll use the same hyperparameter space as in the previous section.

3. Set up the experiment configurations via the `Experiment` class. Note that there are six parameters for the Metis tuner: `optimize_mode`, `no_resampling`, `no_candidates`, `selection_num_starting_points`, `cold_start_num`, and `exploration_probability`. Please refer to the official documentation page for more information regarding the Metis tuner parameters (`https://nni.readthedocs.io/en/stable/reference/hpo.html#metis-tuner`):

    ```
    experiment = Experiment('local')


    experiment.config.experiment_name = 'nni_sklearn_metis'
    experiment.config.tuner.name = 'MetisTuner'
    experiment.config.tuner.class_args['optimize_mode'] =
    'maximize'


    # Boilerplate code
    # same as previous section
    ```

4. Run the hyperparameter tuning experiment:

    ```
    experiment.run(8080, wait_completion = True, debug =
    False)
    ```

5. Train the model on full training data using the best set of hyperparameters found.

    Get the best set of hyperparameters:

    ```
    non_numeric_mapping = params = {
    'model__criterion': ['gini','entropy'],
    'model__class_weight': ['balanced','balanced_subsample'],
        }
    ```

```
best_trial = sorted(experiment.export_data(),key = lambda
x: x.value, reverse = True)[0]
for key in non_numeric_mapping:
    best_trial.parameter[key] = non_numeric_mapping[key]
[best_trial.parameter[key]]
print(best_trial.parameter)
```

Based on the preceding code, we get the following results:

```
{'model__n_estimators': 122, 'model__criterion': 'gini',
'model__class_weight': 'balanced', 'model__min_samples_
split': 0.00173059072806428}
```

We can now train the model on full training data:

```
from sklearn.base import clone
tuned_pipe = clone(pipe).set_params(**best_trial.
parameter)
# Fit the pipeline on train data
tuned_pipe.fit(X_train_full,y_train)
```

6. Test the final trained model on the test data:

```
y_pred = tuned_pipe.predict(X_test_full)
print(f1_score(y_test, y_pred))
```

Based on the preceding code, we get around 0.590 in the F1-score when testing our final trained Random Forest model with the best set of hyperparameters on the test set.

In this section, we have learned how to implement Metis using NNI with pure Python code. In the next section, we will learn how to implement Simulated Annealing with NNI via pure Python code.

Implementing Simulated Annealing

Simulated Annealing is one of the variants of the Heuristic Search hyperparameter tuning group (see *Chapter 5*) that the NNI package can implement. Let's use the same data, pipeline, and hyperparameter space as in the example in the previous section, to implement Simulated Annealing with NNI using pure Python code.

The following code shows how to implement Simulated Annealing with the NNI package using pure Python code. You can find the more detailed code in the GitHub repository mentioned in the *Technical requirements* section:

1. Prepare the model to be tuned in a script. We'll use the same model.py script as in the *Implementing Grid Search* section.

2. Define the hyperparameter space in the form of a Python dictionary. We'll use the same hyperparameter space as in the *Implementing Grid Search* section.

3. Set up the experiment configurations via the `Experiment` class. Note that there is one parameter for the Simulated Annealing tuner, namely `optimize_mode`:

```
experiment = Experiment('local')


experiment.config.experiment_name = 'nni_sklearn_anneal'
experiment.config.tuner.name = 'Anneal'
experiment.config.tuner.class_args['optimize_mode'] =
'maximize'


# Boilerplate code
experiment.config.trial_command = "python '/mnt/c/Users/
Louis\ Owen/Desktop/Packt/Hyperparameter-Tuning-with-
Python/nni/model.py'"
experiment.config.trial_code_directory = '.'
experiment.config.search_space = hyperparameter_space
experiment.config.max_trial_number = 100
experiment.config.trial_concurrency = 10
experiment.config.max_experiment_duration = '1h'
```

4. Run the hyperparameter tuning experiment:

```
experiment.run(8080, wait_completion = True, debug =
False)
```

5. Train the model on full training data using the best set of hyperparameters found.

    Get the best set of hyperparameters:

```
best_trial = sorted(experiment.export_data(),key = lambda
x: x.value, reverse = True)[0]
print(best_trial.parameter)
```

    Based on the preceding code, we get the following results:

```
{'model__n_estimators': 103, 'model__criterion': 'gini',
'model__class_weight': 'balanced_subsample', 'model__min_
samples_split': 0.0010101249953063539}
```

We can now train the model on full training data:

```
from sklearn.base import clone
tuned_pipe = clone(pipe).set_params(**best_trial.
parameter)
# Fit the pipeline on train data
tuned_pipe.fit(X_train_full,y_train)
```

6. Test the final trained model on the test data:

```
y_pred = tuned_pipe.predict(X_test_full)
print(f1_score(y_test, y_pred))
```

Based on the preceding code, we get around 0.600 in the F1-score when testing our final trained Random Forest model with the best set of hyperparameters on the test set.

In this section, we have learned how to implement Simulated Annealing using NNI with pure Python code. In the next section, we will learn how to implement Hyper Band with NNI via pure Python code.

Implementing Hyper Band

Hyper Band is one of the variants of the Multi-Fidelity Optimization hyperparameter tuning group (see *Chapter 6*) that the NNI package can implement. Let's use the same data, pipeline, and hyperparameter space as in the example in the previous section to implement Hyper Band with NNI using pure Python code.

The following code shows how to implement Hyper Band with the NNI package using pure Python code. You can find the more detailed code in the GitHub repository mentioned in the *Technical requirements* section:

1. Prepare the model to be tuned in a script. Here, we'll use a new script called model_advisor. py. In this script, we utilize the TRIAL_BUDGET value from the output of nni.get_ next_parameter() to update the 'model__n_estimators' hyperparameter.

2. Define the hyperparameter space in the form of a Python dictionary. We'll use a similar hyperparameter space to the *Implementing Grid Search* section but we will remove the 'model__n_estimators' hyperparameter since it will become the budget definition for Hyper Band:

```
hyperparameter_space_advisor = {
    'model__criterion': {'_type': 'choice', '_value':
['gini', 'entropy']},
    'model__class_weight': {'_type': 'choice', '_value':
["balanced","balanced_subsample"]},
    'model__min_samples_split': {'_type': 'uniform', '_
```

```
value': [0, 0.1]},
}
```

3. Set up the experiment configurations via the Experiment class. Note that there are four parameters for the Hyper Band advisor: optimize_mode, R, eta, and exec_mode. Please refer to the official documentation page for more information regarding the Hyper Band advisor parameters (https://nni.readthedocs.io/en/latest/reference/hpo.html#hyperband-tuner):

```
experiment = Experiment('local')

experiment.config.experiment_name = 'nni_sklearn_hyper_
band'
experiment.config.advisor.name = 'Hyperband'
experiment.config.advisor.class_args['optimize_mode'] =
'maximize'
experiment.config.advisor.class_args['R'] = 200
experiment.config.advisor.class_args['eta'] = 3
experiment.config.advisor.class_args['exec_mode'] =
'parallelism'

# Boilerplate code
experiment.config.trial_command = "python '/mnt/c/Users/
Louis\ Owen/Desktop/Packt/Hyperparameter-Tuning-with-
Python/nni/model_advisor.py'"
experiment.config.trial_code_directory = '.'
experiment.config.search_space = hyperparameter_space_
advisor
experiment.config.max_trial_number = 100
experiment.config.trial_concurrency = 10
experiment.config.max_experiment_duration = '1h'
```

4. Run the hyperparameter tuning experiment:

```
experiment.run(8080, wait_completion = True, debug =
False)
```

5. Train the model on full training data using the best set of hyperparameters found.

    Get the best set of hyperparameters:

    ```
    best_trial = sorted(experiment.export_data(),key = lambda
    x: x.value, reverse = True)[0]
    best_trial.parameter['model__n_estimators'] = best_trial.
    parameter['TRIAL_BUDGET'] * 50
    del best_trial.parameter['TRIAL_BUDGET']
    print(best_trial.parameter)
    ```

    Based on the preceding code, we get the following results:

    ```
    {'model__criterion': 'gini', 'model__class_weight':
    'balanced_subsample', 'model__min_samples_split':
    0.001676130360763284, 'model__n_estimators': 100}
    ```

    We can now train the model on full training data:

    ```
    from sklearn.base import clone
    tuned_pipe = clone(pipe).set_params(**best_trial.
    parameter)
    ```

    Fit the pipeline on train data.

    ```
    tuned_pipe.fit(X_train_full,y_train)
    ```

6. Test the final trained model on the test data:

    ```
    y_pred = tuned_pipe.predict(X_test_full)
    print(f1_score(y_test, y_pred))
    ```

    Based on the preceding code, we get around 0.593 in the F1-score when testing our final trained Random Forest model with the best set of hyperparameters on the test set.

In this section, we have learned how to implement Hyper Band using NNI with pure Python code. In the next section, we will learn how to implement Bayesian Optimization Hyper Band with NNI via pure Python code.

Implementing Bayesian Optimization Hyper Band

Bayesian Optimization Hyper Band (BOHB) is one of the variants of the Multi-Fidelity Optimization hyperparameter tuning group (see *Chapter 6*) that the NNI package can implement. Note that to use BOHB in NNI, we need to install additional dependencies using the following command:

```
pip install "nni[BOHB]"
```

Let's use the same data, pipeline, and hyperparameter space as in the example in the previous section to implement BOHB with NNI using pure Python code.

The following code shows how to implement Hyper Band with the NNI package using pure Python code. You can find the more detailed code in the GitHub repository mentioned in the *Technical requirements* section:

1. Prepare the model to be tuned in a script. We'll use the same `model_advisor.py` script as in the previous section.

2. Define the hyperparameter space in the form of a Python dictionary. We'll use the same hyperparameter space as in the previous section.

3. Set up the experiment configurations via the `Experiment` class. Note that there are 11 parameters for the BOHB advisor: `optimize_mode`, `min_budget`, `max_budget`, `eta`, `min_points_in_model`, `top_n_percent`, `num_samples`, `random_fraction`, `bandwidth_factor`, `min_bandwidth`, and `config_space`. Please refer to the official documentation page for more information regarding the Hyper Band advisor parameters (`https://nni.readthedocs.io/en/latest/reference/hpo.html#bohb -tuner`):

```
experiment = Experiment('local')

experiment.config.experiment_name = 'nni_sklearn_bohb'
experiment.config.advisor.name = 'BOHB'
experiment.config.advisor.class_args['optimize_mode'] =
'maximize'
experiment.config.advisor.class_args['max_budget'] = 200
experiment.config.advisor.class_args['min_budget'] = 5
experiment.config.advisor.class_args['eta'] = 3


# Boilerplate code
# same as previous section
```

4. Run the hyperparameter tuning experiment:

```
experiment.run(8080, wait_completion = True, debug =
False)
```

5. Train the model on full training data using the best set of hyperparameters found.

    Get the best set of hyperparameters:

    ```
    best_trial = sorted(experiment.export_data(),key = lambda
    x: x.value, reverse = True)[0]
    best_trial.parameter['model__n_estimators'] = best_trial.
    parameter['TRIAL_BUDGET'] * 50
    del best_trial.parameter['TRIAL_BUDGET']
    print(best_trial.parameter)
    ```

    Based on the preceding code, we get the following results:

    ```
    {'model__class_weight': 'balanced', 'model__criterion':
    'gini', 'model__min_samples_split': 0.000396569883631686,
    'model__n_estimators': 1100}
    ```

    We can now train the model on full training data:

    ```
    from sklearn.base import clone
    tuned_pipe = clone(pipe).set_params(**best_trial.
    parameter)
    # Fit the pipeline on train data
    tuned_pipe.fit(X_train_full,y_train)
    ```

6. Test the final trained model on the test data:

    ```
    y_pred = tuned_pipe.predict(X_test_full)
    print(f1_score(y_test, y_pred))
    ```

    Based on the preceding code, we get around 0.617 in the F1-score when testing our final
    trained Random Forest model with the best set of hyperparameters on the test set.

In this section, we have learned how to implement Bayesian Optimization Hyper Band using NNI with
pure Python code. In the next section, we will learn how to implement Population-Based Training
with NNI via nnictl.

Implementing Population-Based Training

Population-Based Training (**PBT**) is one of the variants of the Heuristic Search hyperparameter
tuning group (see *Chapter 5*) that the NNI package can implement. To show you how to implement
PBT with NNI using pure Python code, let's use the same example provided by the NNI package.
Here, the MNIST dataset and a convolutional neural network model are utilized. We'll use PyTorch
to implement the neural network model. For details of the code example provided by NNI,
please refer to the NNI GitHub repository (https://github.com/microsoft/nni/
tree/1546962f83397710fe095538d052dc74bd981707/examples/trials/
mnist-pbt-tuner-pytorch).

> **MNIST Dataset**
>
> MNIST is a dataset of handwritten digits that have been size-normalized and centered in a fixed-size image. Here, we'll use the MNIST dataset provided directly by the PyTorch package (`https://pytorch.org/vision/stable/generated/torchvision.datasets.MNIST.html#torchvision.datasets.MNIST`).

The following code shows how to implement PBT with the NNI package. Here, we'll use `nnictl` instead of using pure Python code. You can find the more detailed code in the GitHub repository mentioned in the *Technical requirements* section:

1. Prepare the model to be tuned in a script. Here, we'll use the same `mnist.py` script from the NNI GitHub repository. Note that we save the script with a new name: `model_pbt.py`.

2. Define the hyperparameter space in a JSON file called `hyperparameter_space_pbt.json`. Here, we'll use the same `search_space.json` file from the NNI GitHub repository.

3. Set up the experiment configurations via the `config_pbt.yaml` file. Note that there are six parameters for the PBT tuner: `optimize_mode`, `all_checkpoint_dir`, `population_size`, `factor`, `resample_probability`, and `fraction`. Please refer to the official documentation page for more information regarding the PBT tuner parameters (`https://nni.readthedocs.io/en/latest/reference/hpo.html#pbt-tuner`):

```
searchSpaceFile: hyperparameter_space_pbt.json
trialCommand: python '/mnt/c/Users/Louis\ Owen/Desktop/
Packt/Hyperparameter-Tuning-with-Python/nni/model_pbt.py'
trialGpuNumber: 1
trialConcurrency: 10
maxTrialNumber: 100
maxExperimentDuration: 1h
tuner:
  name: PBTTuner
  classArgs:
    optimize_mode: maximize
trainingService:
  platform: local
  useActiveGpu: false
```

4. Run the hyperparameter tuning experiment. We can see the experiment status and various interesting stats via the launched web portal. The following code shows how to run the experiment on port 8080 in `local`, meaning you can open the web portal on `http://localhost:8080`:

```
nnictl create --config config_pbt.yaml --port 8080
```

In this section, we have learned how to implement Population-Based Training with NNI via `nnictl` using the same example as provided in the official documentation of NNI.

Summary

In this chapter, we have learned all the important things about the DEAP and Microsoft NNI packages. We also have learned how to implement various hyperparameter tuning methods with the help of these packages, along with understanding each of the important parameters of the classes and how are they related to the theory that we have learned in the previous chapters. From now on, you should be able to utilize these packages to implement your chosen hyperparameter tuning method, and ultimately, boost the performance of your ML model. Equipped with the knowledge from *Chapters 3 – 6*, you will also be able to debug your code if there are errors or unexpected results, and be able to craft your own experiment configuration to match your specific problem.

In the next chapter, we'll learn about hyperparameters for several popular algorithms. There will be a wide explanation for each of the algorithms, including (but not limited to) the definition of each hyperparameter, what will be impacted when the value of each hyperparameter is changed, and the priority list of hyperparameters based on the impact.

Section 3: Putting Things into Practice

In the final section of the book, as its name suggests, we will learn how to put everything we have learned into practice so that we can have an effective and powerful hyperparameter tuning experiment workflow.

This section includes the following chapters:

- *Chapter 11, Understanding Hyperparameters of Popular Algorithms*
- *Chapter 12, Introducing the Hyperparameter Tuning Decision Map*
- *Chapter 13, Tracking Hyperparameter Tuning Experiments*
- *Chapter 14, Conclusions and Next Steps*

11

Understanding the Hyperparameters of Popular Algorithms

Most **machine learning** (**ML**) algorithms have their own hyperparameters. Knowing how to implement a lot of fancy hyperparameter tuning methods without understanding the hyperparameters of the model is the same as a doctor writing a prescription before diagnosing the patient.

In this chapter, we'll learn about the hyperparameters of several popular ML algorithms. There will be a broad explanation for each of the algorithms, including (but not limited to) the definition of each hyperparameter, what will be impacted when the value of each hyperparameter is changed, and the priority list of hyperparameters based on the impact.

By the end of this chapter, you will understand the important hyperparameters of several popular ML algorithms. Understanding the hyperparameters of ML algorithms is crucial since not all hyperparameters are equally significant when it comes to impacting the model's performance. We do not have to perform hyperparameter tuning on all of the hyperparameters of a model; we just need to focus on the more critical hyperparameters.

In this chapter, we will cover the following main topics:

- Exploring Random Forest hyperparameters
- Exploring XGBoost hyperparameters
- Exploring LightGBM hyperparameters
- Exploring CatBoost hyperparameters
- Exploring SVM hyperparameters
- Exploring artificial neural network hyperparameters

Exploring Random Forest hyperparameters

Random Forest is a tree-based model that is built using a collection of **decision trees**. It is a very powerful ensemble ML model that can be utilized for both classification and regression tasks. The way Random Forest utilizes the collection of decision trees is by performing an ensemble method called **bootstrap aggregation** (**bagging**) with some modifications. To understand how each of the Random Forest's hyperparameters can impact the model's performance, we need to understand how the model works in the first place.

Before discussing how Random Forest ensembles a collection of decision trees, let's discuss how a decision tree works at a high level. A decision tree can be utilized to perform a classification or regression task by constructing a series of decisions (in the form of rules and splitting points) that can be visualized in the form of a tree. These decisions are made by looking through all of the features and the feature values of the given training data. The goal of a decision tree is to have high homogeneity for each of the leaf nodes. Several methods can be used to measure homogeneity; the two most popular methods for classification tasks are to calculate the **Gini** or **Entropy** values, while the most popular method for regression tasks is to calculate the **Mean Squared Error** value.

Random Forest utilizes the bagging method to ensemble the collection of decision trees. Bagging is an ensemble method that works by combining predictions from multiple ML models with the hope of generating a more accurate and robust prediction. In this case, Random Forest combines the prediction outputs from several decision trees so that we are not too focused on the prediction from a single tree. This is because a decision tree is very likely to overfit the training data. However, Random Forest does not just utilize the vanilla bagging ensemble method – it also ensures that it only utilizes prediction outputs from the collection of decision trees that are not highly correlated with each other. How is Random Forest able to do that? Instead of asking each decision tree to look through all the features and their values when choosing the splitting points, Random Forest customizes this procedure so that each decision tree only looks at a random sample of features.

The most popular and well-maintained implementation of Random Forest in Python can be found in the scikit-learn package. It includes implementations for both regression (`RandomForestRegressor`) and classification (`RandomForestClassifier`) tasks. Both implementations have very similar hyperparameters with only a few small differences. The following are the most important hyperparameters, starting with the most important to the least based on the impact on model performance. Note that this priority list is subjective, based on our experience of developing Random Forest models in the past:

1. `n_estimators`: This specifies the number of decision trees to be utilized to build the Random Forest. In general, the larger the number of trees, the better the model's performance will be, with a trade-off of having longer computational time. However, there is a threshold beyond which adding more trees will not have much additional impact on the model's performance. It could even have a negative impact due to the problem of overfitting.

2.   `max_features`: This specifies the number of randomly sampled features that are used by Random Forest to choose the best splitting point in each of the decision trees. The higher the value, the lower the reduction in variance, and hence the lower the increase in bias. A higher value also leads to having a longer computational time. scikit-learn, by default, will use all of the features for regression tasks and use only `sqrt(n_features)` number of features for classification tasks.

3.   `criterion`: This is used to measure the homogeneity of each decision tree. scikit-learn implemented several methods for both regression and classification tasks. There's `squared_error`, `absolute_error`, and `poisson` for regression tasks, while there's `gini`, `entropy`, and `log_loss` for classification tasks. Different methods will have different impacts on model performance; there is no clear rule of thumb for this hyperparameter.

4.   `max_depth`: This specifies the maximum depth of each decision tree. The default value of this hyperparameter is `None`, meaning that the nodes of each tree will keep branching until we have pure leaf nodes or until all the leaves contain less than `min_samples_split` number of samples. The lower the value, the better, since this prevents overfitting. However, a value that is too low can lead to an underfitting problem. One thing is for sure – a higher value implies a longer computational time.

5.   `min_samples_split`: This specifies the minimum number of samples required for a tree to be able to further split an internal node (a node that can be split into child nodes). The higher the value, the easier it is to prevent overfitting.

6.   `min_samples_leaf`: This specifies the minimum number of samples required in the leaf nodes. A higher value can help us prevent overfitting.

> **Random Forest Hyperparameters in scikit-learn**
>
> For more information about each of the hyperparameters of the Random Forest implementation in scikit-learn, please visit the official documentation pages at `https://scikit-learn.org/stable/modules/generated/sklearn.ensemble.RandomForestClassifier.html` and `https://scikit-learn.org/stable/modules/generated/sklearn.ensemble.RandomForestRegressor.html`.

Other useful boilerplate parameters can be found across different scikit-learn estimator implementations. The following are several important parameters that you need to be aware of that can help you while training a scikit-learn estimator:

1.   `class_weight`: This specifies the weights for each class that exists in the training data. This is only available for classification tasks. This parameter is very important when you face an imbalanced class problem. We need to give higher weights to classes that have fewer samples.

2.   `n_jobs`: This specifies the number of parallel processes to be utilized when training the estimator. scikit-learn utilizes the `joblib` package in the backend.

3. `random_state`: This specifies the random seed number to ensure the code is reproducible.

4. `verbose`: This parameter is used to control any logging activities. Setting `verbose` to an integer greater than zero enables us to see what happens when training an estimator.

In this section, we learned how Random Forest works at a high level and looked at several important hyperparameters, along with an explanation of how they impact the model's performance. We also looked at the main hyperparameters. Furthermore, we learned about several useful parameters in scikit-learn that can ease the training process. In the next section, we will discuss the XGBoost algorithm.

Exploring XGBoost hyperparameters

Extreme Gradient Boosting (**XGBoost**) is also a tree-based model that is built using a collection of decision trees, similar to a Random Forest. It can also be utilized for both classification and regression tasks. The difference between XGBoost and Random Forest is in how they perform the ensemble. Unlike Random Forest, which uses the bagging ensemble method, XGBoost utilizes another ensemble method called **boosting**.

Boosting is an ensemble algorithm whose goal is to achieve higher performance through a sequence of individually weak models by overcoming the weaknesses of the predecessor models (see *Figure 11.1*). It is not a specific model; it's just a generic ensemble algorithm. The definition of weakness may vary across different types of boosting ensemble implementation. In XGBoost, it is defined based on the error of the gradient from the previous decision tree model. Take a look at the following diagram:

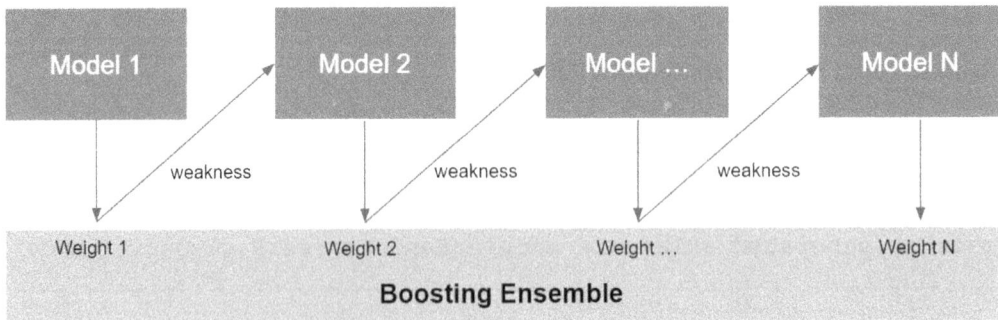

Figure 11.1 – Boosting ensemble algorithm

XGBoost is a very popular and well-adopted ML model that is built using the boosting ensemble algorithm and a collection of decision trees. Each of the decision trees is added one at a time and is fitted to the prediction errors from the previous tree to correct those errors. It is worth noting that since XGBoost is part of the gradient boosting algorithm, all of the weak models (decision trees) need to be fitted using a differentiable loss function and the gradient descent optimization method.

XGBoost has its own package and can be utilized not only in Python but also in other programming languages, such as R and JVM. In Python, you can install XGBoost via `pip install xgboost`. This package also implements the scikit-learn wrappers for both regression (`XGBRegressor`) and classification (`XGBClassifier`) tasks. Numerous hyperparameters are provided by the package, but not all of them are very important in affecting the model's performance. The following are the most important hyperparameters, starting with the most important to the least based on their impact on model performance:

1. `n_estimators`: This specifies the number of decision trees to be utilized to build the XGBoost model. It can also be interpreted as the number of boosting rounds, which is similar to the concept of epochs in a neural network. In general, the higher the value, the better the model's performance will be, with the trade-off of having a longer computation time. However, we need to be careful with a value that's too high since it can lead us to the overfitting problem.

2. `learning_rate`: This is the learning rate of the gradient descent optimization algorithm. The lower the value, the higher the chances of the model finding the optimum solution, with a trade-off of having a longer computational time. You can increase the value of this hyperparameter if there no sign of overfitting is found on the last iterations of training; you can decrease it if there is overfitting.

3. `max_depth`: This is the maximum depth of each decision tree. A lower value can help us prevent overfitting. However, a too-low value can lead to an underfitting problem. One thing is for sure – a higher value leads to a longer computational time.

4. `min_child_weight`: This is the minimum sum of instance weight, calculated using the Hessian, that's needed in a child. This hyperparameter acts as a regularizer to ensure that each tree will stop trying to split the node once a certain degree of purity is reached. In other words, it is a regularization parameter that works by limiting the depth of the tree so that the overfitting problem can be prevented. A higher value can help us prevent overfitting. However, a too-high value can lead to an underfitting problem.

5. `gamma`: This is a pseudo-regularization parameter that is calculated based on a reduction in the loss value. The value of this hyperparameter specifies the minimum loss reduction required to make a further partition on a leaf node of the tree. You can put a high value on this hyperparameter to prevent the overfitting problem. However, please be careful and don't use a value that's too high; it can lead to an underfitting problem.

6. `colsample_bytree`: This is the fraction version of the `max_features` hyperparameter in the scikit-learn implementation of Random Forest. This hyperparameter is responsible for telling XGBoost how many randomly sampled features are needed to choose the best splitting point in each of the decision trees. A low value can help us prevent overfitting and lowers the computational time. However, a value that's too low can lead to an underfitting problem.

7. `subsample`: This is the observation's version of the `colsample_bytree` hyperparameter. This hyperparameter is responsible for telling XGBoost how many training samples need to be used while training each tree. This hyperparameter can be useful to prevent the overfitting problem. However, it can also lead us to an underfitting problem if we use a value that's too low.

> **Complete List of XGBoost Hyperparameters**
>
> For more information about other XGBoost's hyperparameters, please visit the official documentation page: `https://xgboost.readthedocs.io/en/stable/python/python_api.html#module-xgboost.sklearn`.

In this section, we discussed how XGBoost works at a high level and looked at several important hyperparameters, along with an explanation of how they impact model performance. We also looked at the main hyperparameters. In the next section, we will discuss the LightGBM algorithm.

Exploring LightGBM hyperparameters

Light Gradient Boosting Machine (**LightGBM**) is also a boosting algorithm built on top of a collection of decision trees, similar to XGBoost. It can also be utilized both for classification and regression tasks. However, it differs from XGBoost in the way the trees are grown. In LightGBM, trees are grown in a leaf-wise manner, while XGBoost grows trees in a level-wise manner (see *Figure 11.2*). By leaf-wise, we mean that LightGBM grows trees by prioritizing nodes whose split leads to the highest increase of homogeneity:

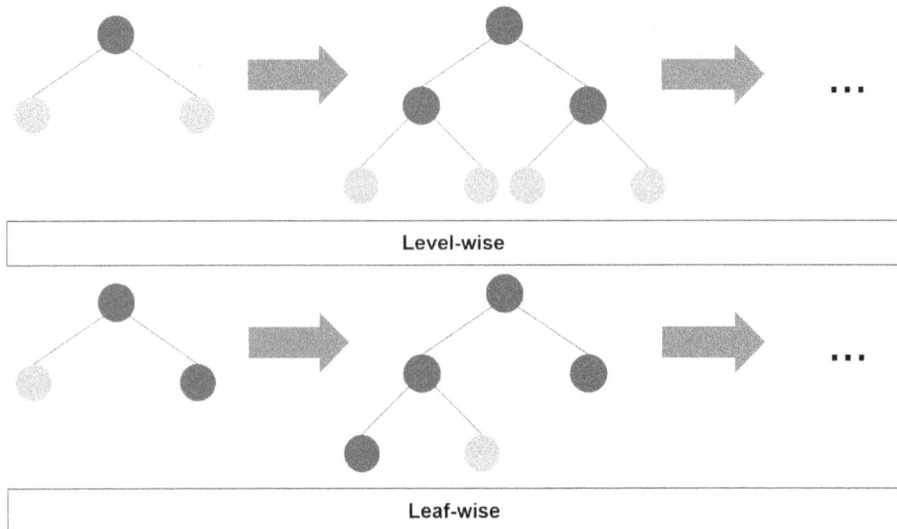

Figure 11.2 – Level-wise versus leaf-wise tree growth

Besides the difference in how XGBoost and LightGBM grow the trees, they also have different ways of handling categorical features. In XGBoost, we need to encode the categorical features before passing them to the model. This is usually done using the one-hot encoding or integer encoding methods. In LightGBM, we can just tell which features are categorical and it will handle those features automatically by performing equality splitting. There are several other differences between XGBoost and LightGBM in terms of the way they perform optimization in distributed learning. In general, LightGBM has a much faster computation time compared to XGBoost.

Similar to XGBoost, LightGBM also has its own package and can be utilized not only in Python but also in the R language. In Python, you can install LightGBM via `pip install lightgbm`. This package also implements the scikit-learn wrappers for both regression (`LGBMRegressor`) and classification (`LGBMClassifier`) tasks. The following are the most important hyperparameters for LightGBM, starting with the most important to the least based on the impact on model performance:

1. `max_depth`: This specifies the maximum depth of each decision tree. A lower value can help us prevent overfitting. However, a value that's too low can lead to an underfitting problem. One thing is for sure – a higher value implies a longer computational time.

2. `num_leaves`: This specifies the maximum number of leaves in each tree. It should have a value lower than two to the power of `max_depth` since a leaf-wise tree is much deeper than a depth-wise tree for a set number of leaves. In general, the higher the value, the better the model's performance will be, with a trade-off of having a longer computational time. However, there is a threshold where the impact of adding more leaves will not have much additional impact on the model's performance or even have a negative impact due to overfitting.

3. `Learning_rate`: This specifies the learning rate of the gradient descent optimization algorithm. The lower the value, the higher the chances of the model finding a more optimum solution, with a trade-off of having a longer computational time. You can increase the value of this hyperparameter if no sign of overfitting is found on the last iterations of training and vice versa.

4. `min_child_samples`: This specifies the minimum number of samples required in the leaf nodes. A higher value can help us prevent overfitting. However, a value that's too high can lead to an underfitting problem.

5. `Feature_fraction`: This is similar to `colsample_bytree` in XGBoost. This hyperparameter tells LightGBM how many randomly sampled features need to be used to choose the best splitting point in each of the decision trees. This hyperparameter can be useful for preventing overfitting. However, it can also lead to an underfitting problem if we use a value that is too low.

6. `bagging_fraction`: This is the observation's version of the `feature_fraction` hyperparameter. This hyperparameter is responsible for telling LightGBM how many training samples need to be used during the training of each tree. Lower values can help us prevent overfitting and lower the computational time. However, a value that is too low can lead to an underfitting problem.

> **Complete List of LightGBM Hyperparameters**
>
> For more information about other LightGBM hyperparameters, please visit the official documentation page: `https://lightgbm.readthedocs.io/en/latest/Python-API.html#scikit-learn-api`.

In this section, we discussed how LightGBM works at a high level and looked at several important hyperparameters, along with an explanation of how they impact model performance. We also looked at the main hyperparameters. In the next section, we will discuss the CatBoost algorithm.

Exploring CatBoost hyperparameters

Categorical Boosting (**CatBoost**) is another boosting algorithm built on top of a collection of decision trees, similar to XGBoost and LightGBM. It can also be utilized both for classification and regression tasks. The main difference between CatBoost and XGBoost or LightGBM is how it grows the trees. In XGBoost and LightGBM, trees are grown asymmetrically, while in CatBoost, trees are grown symmetrically so that all of the trees are balanced. This balanced tree characteristic provides several benefits, including the ability to control overfitting problems, lower inference time, and efficient implementation in CPUs. CatBoost does this by using the same condition in every split in the nodes, as shown in the following diagram:

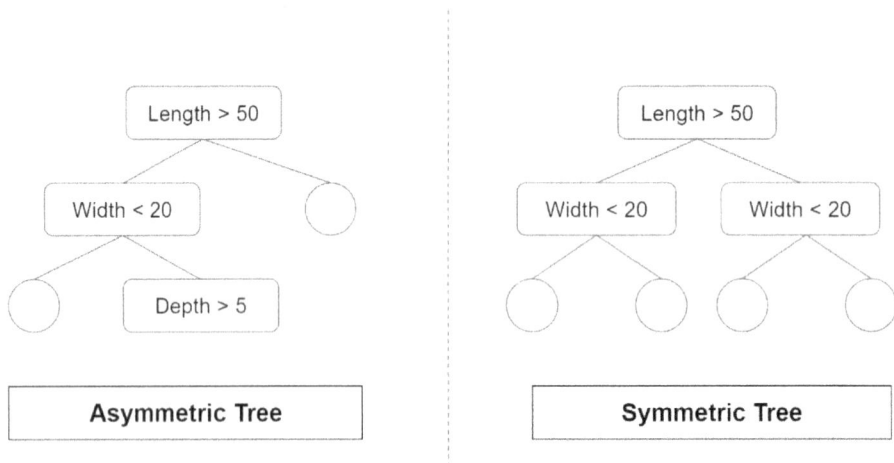

Figure 11.3 – Asymmetric versus symmetric tree

The main selling point of CatBoost is its ability to handle numerous types of features automatically, including numerical, categorical, and text, especially for categorical features. We just need to tell CatBoost which features are categorical features via the `cat_features` parameter and it will handle those features automatically. By default, CatBoost will perform one-hot encoding for categorical features that only have two classes. For higher cardinality features, it will perform target encoding and combine several categorical features or even categorical and numerical features. For more information on how CatBoost handles categorical features, please refer to the official documentation page: `https://catboost.ai/en/docs/concepts/algorithm-main-stages_cat-to-numberic`.

Similar to XGBoost and LightGBM, CatBoost also has its own package and can be utilized not only in Python but also in the R language. In Python, you can install CatBoost via `pip install catboost`. You can utilize the implemented scikit-learn-friendly classes for both regression (`CatBoostRegressor`) and classification (`CatBoostClassifier`) tasks. The following is a list of CatBoost's most important hyperparameters, sorted in descending order based on the importance of each hyperparameter regarding model performance:

1. `iterations`: This specifies the number of decision trees to be utilized to build the CatBoost model. It can also be interpreted as the number of boosting rounds, similar to the concept of epochs in a neural network. In general, the higher the value, the better the model's performance will be, with a trade-off of having a longer computational time. However, there is a threshold where the impact of adding more trees will not have much additional impact on the model's performance or even have a negative impact due to overfitting.

2. `depth`: This specifies the maximum depth of each decision tree. A lower value can help us prevent overfitting. However, a value that's too low can lead to an underfitting problem. One thing is for sure – a higher value implies a longer computational time.

3. `learning_rate`: This specifies the learning rate of the gradient descent optimization algorithm. The lower the value, the higher the chances of the model finding a more optimum solution, with a trade-off of having a longer computational time. You can increase the value of this hyperparameter if no sign of overfitting is found on the last iterations of training and vice versa.

4. `l2_leaf_reg`: This is the regularization parameter on the cost function. This hyperparameter can prevent the overfitting problem. However, it can also lead to an underfitting problem if we use a value that's too high.

5. `one_hot_max_size`: This is the threshold that tells CatBoost when to perform one-hot encoding on the categorical features. Any categorical features that have cardinality lower than or equal to the given value will be transformed into numerical values via the one-hot encoding method.

Complete List of CatBoost Hyperparameters

For more information about other CatBoost hyperparameters, please visit the official documentation page (`https://catboost.ai/en/docs/concepts/parameter-tuning`).

In this section, we discussed how CatBoost works at a high level and looked at several important hyperparameters, along with an explanation of how they impact model performance. We also looked at the main hyperparameters. In the next section, we will discuss the SVM algorithm.

Exploring SVM hyperparameters

Support Vector Machine (**SVM**) is an ML model that utilizes lines or hyperplanes, along with some linear algebra transformations, to perform a classification or regression task. All the algorithms discussed in the previous sections can be classified as tree-based algorithms, while SVM is not part of the *tree-based* group of ML algorithms. It is part of the *distance-based* group of algorithms. We usually called the linear algebra transformation in SVM a **kernel**. This is responsible for transforming any problem into a linear problem.

The most popular and well-maintained implementation of SVM in Python can be found in the scikit-learn package. It includes implementations for both regression (SVR) and classification (SVC) tasks. Both of them have very similar hyperparameters with only a few small differences. The following are the most important hyperparameters for SVM, starting with the most important to the least based on their impact on model performance:

1. kernel: This is the linear algebra transformation, whose goal is to convert the given problem into a linear problem. There are five kernels that we can choose from, including linear (linear), polynomial (poly), radial basis function (rbf), and sigmoid (sigmoid) kernels. Different kernels will have different impacts on model performance and there is no clear rule of thumb for this hyperparameter.

2. C: This is the regularization parameter that controls overfitting. The lower the value, the stronger the impact that regularization will have on the model, and hence a higher chance of preventing overfitting.

3. degree: This hyperparameter is specific to the polynomial kernel function. The value of this hyperparameter corresponds to the degree of the polynomial function that's used by the model.

4. gamma: This is the coefficient for the radial basis, polynomial, and sigmoid kernel functions. There are two options that scikit-learn provides, namely scale and auto.

SVM Hyperparameters in scikit-learn

For more information about how each of the hyperparameters in SVM are implemented in scikit-learn, you can visit the official documentation pages at https://scikit-learn. org/stable/modules/generated/sklearn.svm.SVC.html and https:// scikit-learn.org/stable/modules/generated/sklearn.svm.SVR.html.

In this section, we discussed how SVM works at a high level and looked at several important hyperparameters, along with an explanation of how they impact model performance. We also looked at the main hyperparameters. In the next section, we will discuss artificial neural networks.

Exploring artificial neural network hyperparameters

An **artificial neural network**, also known as **deep learning**, is a kind of ML algorithm that mimics how human brains work. Deep learning can be utilized for both regression and classification tasks. One of the main selling points of this model is its ability to perform feature engineering and selection automatically from the raw data. In general, to ensure this algorithm works decently, we need a large amount of training data to be fed to the model. The simplest form of a neural network is called a **perceptron** (see *Figure 11.4*). A perceptron is just a linear combination that is applied on top of all of the features, with bias added at the end of the calculation:

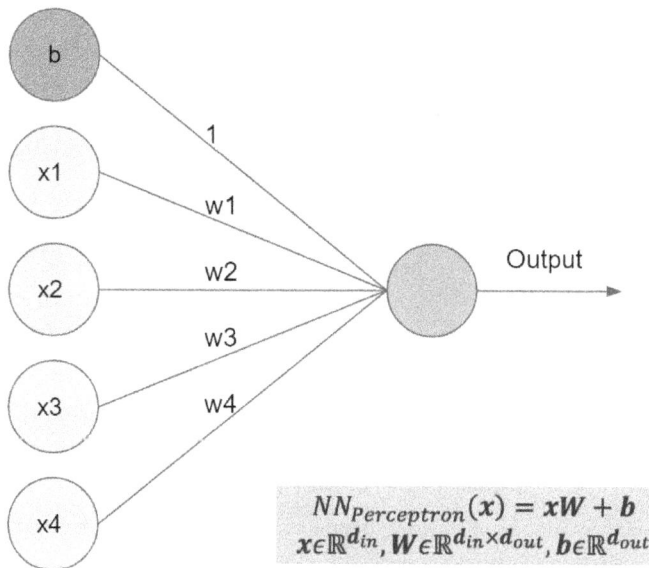

$$NN_{Perceptron}(x) = xW + b$$
$$x \in \mathbb{R}^{d_{in}}, W \in \mathbb{R}^{d_{in} \times d_{out}}, b \in \mathbb{R}^{d_{out}}$$

Figure 11.4 – Perceptron

If the output from the perceptron is passed to a non-linear function, which is usually called an **activation function**, and then passed to another perceptron, then we can call this a **multi-layer perceptron** (**MLP**) with one layer. The training process for a neural network consists of two big procedures, namely **forward propagation** and **backward propagation**. In forward propagation, we just let the neural network perform calculations on top of the given inputs based on the defined architecture. In backward propagation, the model will update the weights and bias parameters based on the defined loss function using a gradient-based optimization procedure.

There are other variants of neural networks besides MLP, such as **convolutional neural networks (CNNs)**, **long short-term memory networks (LSTMs)**, **recurrent neural networks (RNNs)**, and **transformers**. CNN is usually adopted when we work with image data, but we can also use a one-dimensional CNN when working with text data. RNNs and LSTMs are usually adopted when working with time series or natural language data. Transformers are mainly used for text-related projects, but recently, they have been adopted for image and voice data.

Several packages provide implementations of neural networks in Python, including PyTorch, TensorFlow, and Scikit Learn. The following are the most important hyperparameters, sorted in descending order based on the importance of each hyperparameter regarding model performance. Note that this priority list is subjective based on our experience of developing Random Forest models in the past. Since the naming of the hyperparameters may differ across different packages, we will only use the general names of the hyperparameters:

1. **Optimizer**: This is the gradient-based optimization algorithm to be used. There are several optimizers for us to choose from. However, perhaps the most popular and widely adopted optimizer is **Adam**. There are other options, including (but not limited to) SGD and RMSProp. Different optimizers may have different impacts on model performance and there is no clear rule of thumb for choosing which one is the best. It is worth noting that each optimizer has its own hyperparameter as well.

2. **Learning Rate**: This hyperparameter controls how big the step will be for the optimizer to "learn" from the given training data during the optimization process. It is important to choose the best range of learning rates first before tuning other hyperparameters. The lower the value, the higher the chances of the model finding a more optimum solution, with a trade-off of having a longer computational time. You can increase the value of this hyperparameter if no sign of overfitting is found on the last iterations of training and vice versa.

3. **Batch Size**: This specifies the number of training samples that will be passed to the neural network within each training step. In general, the higher the value, the better the model's performance will be. However, a batch size that's too high will usually be constrained by the device's memory.

4. **Epochs**: This is the number of training iterations. Similar to `n_estimators` in XGBoost and `iterations` in CatBoost, a high value can lead to better model performance, with a trade-off of having a longer computational time. However, we need to be careful when using a value that's too high since it can lead to overfitting.

5. **Number of Layers**: The higher the value, the higher the complexity of the model, hence the higher the chance of overfitting. Usually, one or two layers is more than enough to build a good model.

6. **Number of Nodes**: The number of units or nodes within each of the layers. The higher the value, the higher the complexity of the model, hence a higher chance of overfitting.

7. **Activation Function**: The non-linear transformation function. There are many activation functions to choose from. Some of the most well-adopted activation functions in practice are **Rectified Linear Activation Function (ReLU)**, **Exponential Linear Unit (ELU)**, **Sigmoid**, **Softmax**, and **Tanh**.

8. **Dropout Rate**: The rate for the dropout layer. The dropout layer is a special layer in a neural network that acts as a regularizer by randomly setting the unit value to zero. This hyperparameter controls how many units are set to zero. A higher value can help us prevent overfitting. However, a value that's too high can lead to an underfitting problem.

9. **L1/L2 Regularization**: These are the regularization parameters that are applied to the loss function. This hyperparameter can help prevent overfitting. However, it can also lead to an underfitting problem if its value is too high.

In this section, we have discussed how neural network works at a high level, the variants of neural networks, and looked at several important hyperparameters, along with an explanation of how they impact model performance. We also looked at the main hyperparameters. Now, let's summarize this chapter.

Summary

In this chapter, we discussed how several popular algorithms work at a high level, explained their important hyperparameters and how they impact performance, and provided priority lists of the hyperparameters, sorted in descending order based on their impact on performance. At this point, you should be able to design your hyperparameter tuning experiments more effectively by focusing on the most important hyperparameters. You should also understand what impact each of the important hyperparameters has on the performance of the model.

In the next chapter, we'll summarize the hyperparameter tuning methods we've discussed here into a simple decision map that can help you choose which method is the most suitable for your problem. Furthermore, we will cover several study cases that show how to utilize this decision map in practice.

12

Introducing Hyperparameter Tuning Decision Map

Getting too much information can sometimes lead to confusion, which can, in turn, lead back to adopting the simplest option. We learned about numerous hyperparameter tuning methods in the previous chapters. Although we have discussed the ins and outs of each method, it will be very useful for us to have a single source of truth that can be used to help us decide which method to use in which situation.

In this chapter, you'll be introduced to the **Hyperparameter Tuning Decision Map** (**HTDM**), which summarizes all of the discussed hyperparameter tuning methods into a simple decision map based on many aspects, including the properties of the hyperparameter space, the complexity of the objective function, training data size, computational resources availability, prior knowledge availability, and the types of ML algorithms we are working with. There will be also three study cases that show how to utilize HTDM in practice.

By the end of this chapter, you'll be able to utilize HTDM in practice to help you decide which hyperparameter tuning method to be adopted in your specific situation.

In this chapter, we will cover the following main topics:

- Getting familiar with HTDM
- Case study 1 – using HTDM with a CatBoost classifier
- Case study 2 – using HTDM with a conditional hyperparameter space
- Case study 3 – using HTDM with prior knowledge of the hyperparameter values

Getting familiar with HTDM

HTDM is designed to help you decide which hyperparameter tuning method should be adopted in a particular situation (see *Figure 12.1*). Here, the situation is defined based on six aspects:

- Hyperparameter space properties, including the size of the space, types of hyperparameter values (numerical only or mixed), and whether it contains conditional hyperparameters or not

- Objective function complexity: whether it is a cheap or expensive objective function

- Computational resource availability: whether or not you have enough parallel computational resources

- Training data size: whether you have a few, moderate, or a large number of training samples

- Prior knowledge availability: whether you have prior knowledge of the good range of hyperparameter values

- Types of ML algorithms: whether you are working with a small, medium, or large-sized model, and whether you are working with a traditional ML or deep learning type of algorithm

This can be seen in the following diagram:

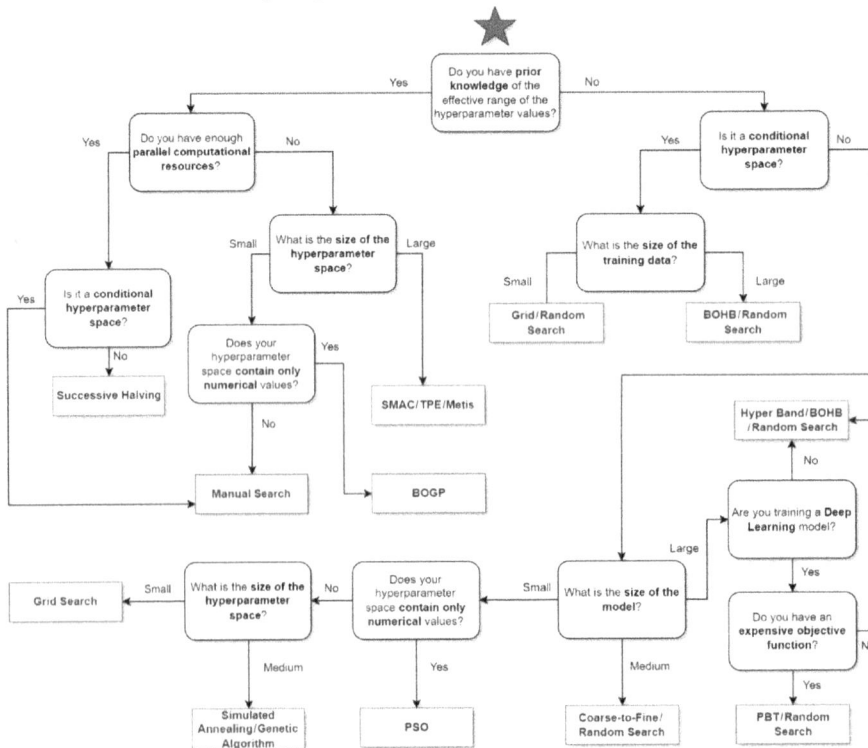

Figure 12.1 – HTDM

The definition of *Small*, *Medium*, and *Large* in HTDM is very subjective. However, you can refer to the following table as a rule of thumb:

Aspect	Size	Rule of Thumb
Training data size	Small	< 10,000 samples
	Large	>= 10,000 samples
Hyperparameter space size	Small	< 5 hyperparameters
	Medium	< 10 hyperparameters
	Large	>= 10 hyperparameters
ML model size	Small	< 50 MB
	Medium	< 300 MB
	Large	>= 300 MB

Figure 12.2 – Rule of thumb of size definition

The following important notes may also help us decide which hyperparameter tuning method we should adopt in a particular situation:

Group	Method	Notes	Supported Packages
Exhaustive Search	Manual Search	Best combined with other methods.	
	Grid Search	Knows exactly what the important hyperparameters to be tuned are. Wants to get the exact optimal solution.	scikit-learn, Optuna, NNI
	Random Search	Works better than grid search when there's an unimportant hyperparameter in the space.	scikit-learn, Hyperopt, Optuna, NNI
Bayesian Optimization	BOGP	There are variants of BOGP that also support non-numerical types of hyperparameters.	scikit-learn, NNI
	SMAC	Best when the hyperparameter space is dominated by categorical hyperparameters.	NNI
	TPE	Unlike SMAC, TPE doesn't focus only on the best-observed points during the trials but the distribution of the best-observed points.	Hyperopt, Optuna, NNI
	Metis	Similar to SMAC and TPE, but also wants to know which set of hyperparameters should be tested in the next trial.	NNI

Group	Method	Notes	Supported Packages
Heuristic Search	Simulated Annealing	May skip parts of the search space that contain optimal hyperparameters.	Hyperopt, Optuna, NNI
	Genetic Algorithm	High computation cost due to the need to evaluate all individuals in each generation.	DEAP
	PSO	Works well only with continuous types of hyperparameters, but can be modified for discrete hyperparameters as well.	DEAP
	PBT	Just wants the final trained model without knowing the hyperparameter configuration.	NNI
Multi-Fidelity Optimization	Coarse-to-Fine	A very simple method with a customizable module based on your preference.	scikit-learn
	Successive Halving	Has metadata or previous experience setting up the trade-off between the number of resources and candidates to be sampled.	scikit-learn, Optuna
	Hyper Band	Similar to Successive Halving, but does not have time or metadata to help you configure the trade-off between the number of resources and candidates.	scikit-learn, Optuna, NNI
	BOHB	Can decide which subspace needs to be searched based on previous experiences, not based on luck. Not only can it get a strong initial performance (inherited from HB), but it can also get a strong final performance (inherited from BO).	NNI

Figure 12.3 – Important notes for each hyperparameter tuning method

In this section, we discussed HTDM, along with several additional important notes to help you decide which hyperparameter tuning method you should adopt in a particular situation. In the next few sections, we will learn how to utilize HTDM in practice through several interesting study cases.

Case study 1 – using HTDM with a CatBoost classifier

Let's say we are training a classifier based on the marketing campaign data that was introduced in *Chapter 7, Hyperparameter Tuning via scikit*. Here, we are utilizing CatBoost (see *Chapter 11, Understanding Hyperparameters of Popular Algorithms*) as the classifier. This is our first time working with the given data. The laptop we are using only has a single-core CPU and the hyperparameter space is defined as follows. Note that we are not working with a conditional hyperparameter space:

- `iterations: randint(5,200)`
- `depth: randint(3,10)`
- `learning_rate: np.linspace(1e-5,1e-3,20)`
- `l2_leaf_reg: np.linspace(1,30,30)`
- `one_hot_max_size: randint(2,15)`

Based on the given case description, we can try to utilize HTDM to help us choose which hyperparameter tuning suits the condition the best. First of all, we know that we do not have any prior knowledge or meta-learning results of the good hyperparameter values on the given data. This means we will only focus on the right branch of the first node in HTDM, as shown here:

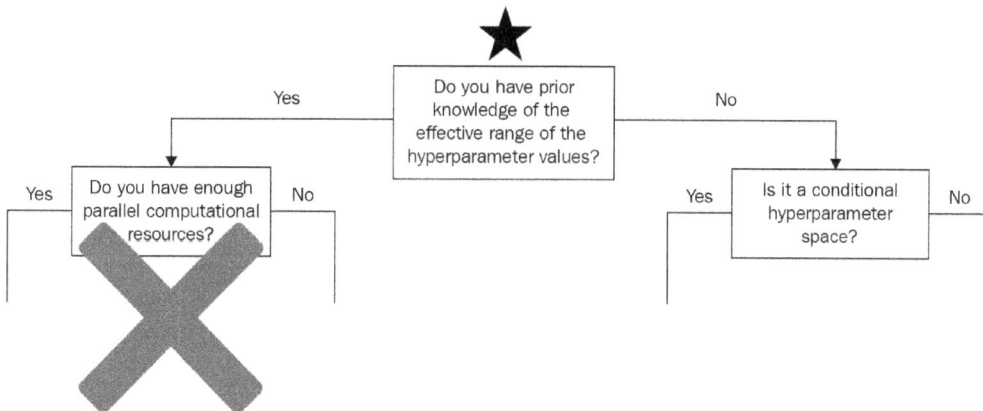

Figure 12.4 – Case study 1, no prior knowledge

We know that we are not working with a conditional hyperparameter space. This means we will only focus on the right branch of the second node, as shown here:

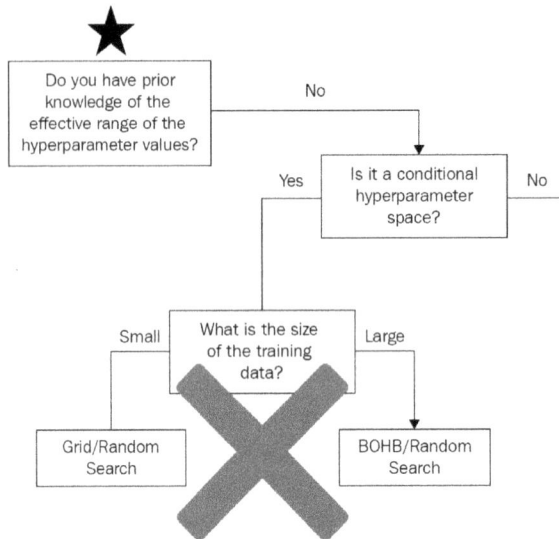

Figure 12.5 – Case study 1, not a conditional hyperparameter space

Based on a rough estimation, our CatBoost model's size should be in the range of small to medium-sized. This means we will only focus on the left and bottom branches of the third node, as shown here:

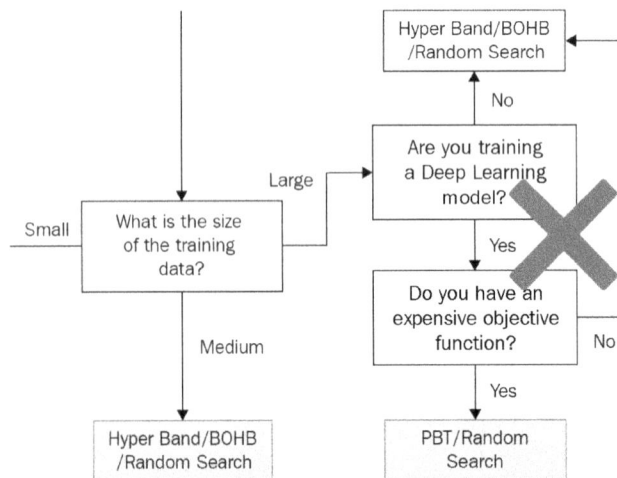

Figure 12.6 – Case study 1, small to medium model size

We also have a medium-sized hyperparameter space that consists of only numerical values. This means our options are Coarse-to-Fine, Random Search, PSO, Simulated Annealing, and Genetic Algorithm. It is worth noting that even though our hyperparameter space consists of only numerical values, we can still utilize hyperparameter tuning methods that work with mixed types of values:

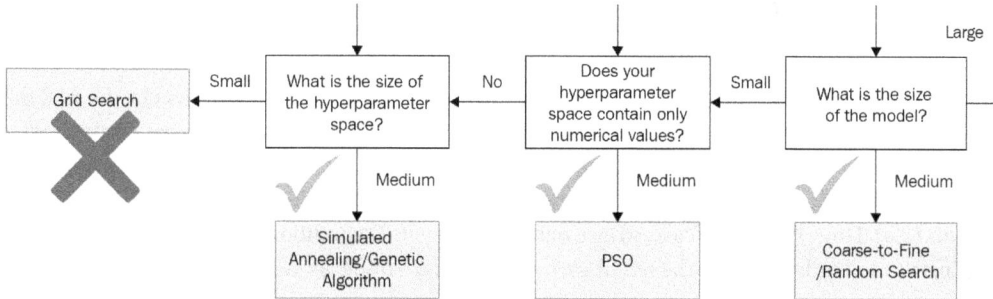

Figure 12.7 – Case study 1, medium-sized hyperparameter space with only numerical values

So, how do we choose a hyperparameter tuning method from the selected options? First, we know that PSO only works very well on the continuous type of hyperparameter values while we also have integers in the hyperparameter space. Thus, we can remove PSO from our options. This leaves us with the remaining four options. One easy and effective way to choose the best hyperparameter tuning method is by choosing the simplest method, which is the Random Search method.

In this section, we discussed the first case study on how to utilize HTDM in practice. In the next section, we will do the same using another interesting case study.

Case study 2 – using HTDM with a conditional hyperparameter space

Let's say we are faced with a similar condition as in the previous section but now, we are working with a conditional hyperparameter space, as defined here:

```
one_hot_max_size = randint(2,15)
iterations = randint(5,200)
If iterations < 50:
    depth = randint(3,10)
    learning_rate = np.linspace(5e-4,1e-3,10)
    l2_leaf_reg = np.linspace(1,15,20)
elif iterations < 100:
    depth = randint(3,7)
```

```
    learning_rate = np.linspace(1e-5,5e-4,10)
l2_leaf_reg = np.linspace(5,20,20)
else:
    depth = randint(3,5)
    learning_rate = np.linspace(1e-6,5e-5,10)
l2_leaf_reg = np.linspace(5,30,20)
```

Based on the given case description, we can try to utilize HTDM again to help us choose which hyperparameter tuning method suits the condition the best. Here, similar to the previous study case, we know that we do not have any prior knowledge or meta-learning results of the good hyperparameter values on the given data. This means we will only focus on the right branch of the first node in HTDM (see *Figure 12.4*). However, in this case, we are now working with a conditional hyperparameter space. This means we will only focus on the left branch of the second node, as shown here:

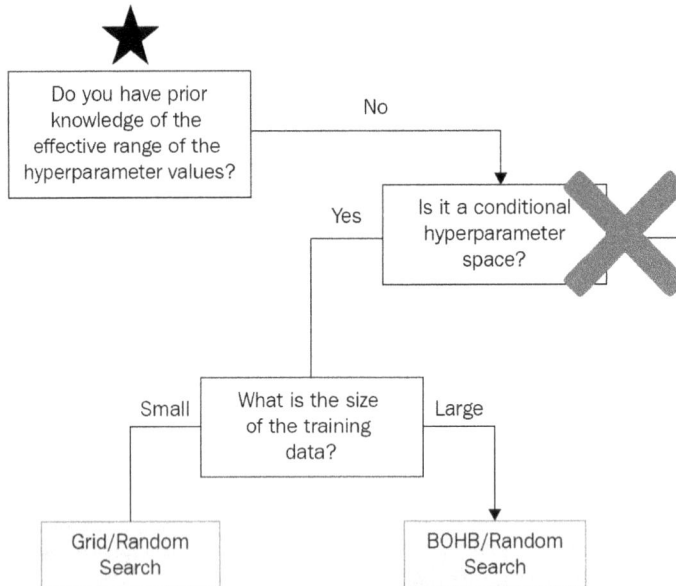

Figure 12.8 – Case study 2, a conditional hyperparameter space

Since we have more than 10,000 samples of training data (see *Chapter 7, Hyperparameter Tuning via scikit*), we only have two hyperparameter tuning methods to choose from based on HTDM, namely the BOHB or Random Search method (see *Figure 12.9*). Choosing Random Search over BOHB surely is a wise choice if we only compare them based on the simplicity of the implementation since we need to install the Microsoft NNI package just to adopt the BOHB method (see *Figure 12.3*).

However, we know that we are working with a model that is not very small, and BOHB can decide which subspace needs to be searched based on previous experiences, not based on luck. Thus, in theory, BOHB will be a better choice to save us time searching for the best set of hyperparameters. So, which method should we pick? It's up to your discretion:

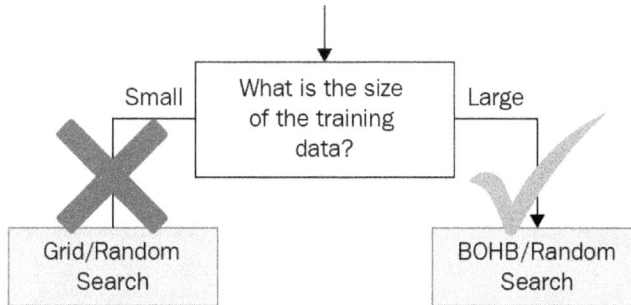

Figure 12.9 – Case study 2, large training data

In this section, we discussed the second case study on how to utilize HTDM in practice. In the next section, we will do the same using another interesting case study.

Case study 3 – using HTDM with prior knowledge of the hyperparameter values

Let's say, in this case, we are also faced with a similar condition as in the previous case study, but this time, we have prior knowledge of the good hyperparameter values for the given data since one of the data scientists in our team has worked with the same data previously. This means we will only focus on the left branch of the first node in HTDM, as shown here:

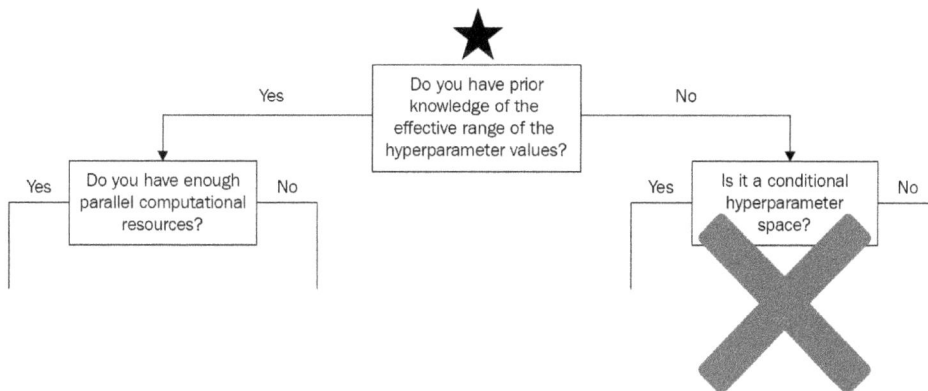

Figure 12.10 – Case study 3, have prior knowledge

Based on the given case description, we know that we do not have enough parallel computational resources since we only have a single-core CPU. This means we will only focus on the right branch of the second node, as shown here:

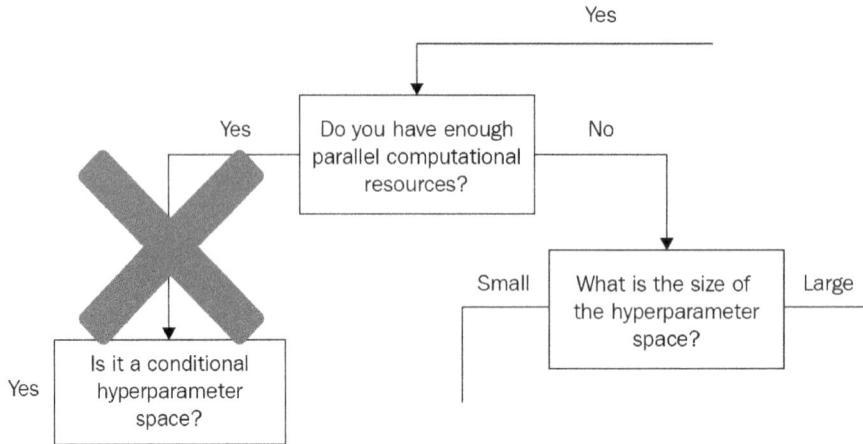

Figure 12.11 – Case study 3, not enough parallel computational resources

We also know that we have a medium-sized hyperparameter space that only consists of numerical types of values. This means our options are SMAC, TPE, and Metis:

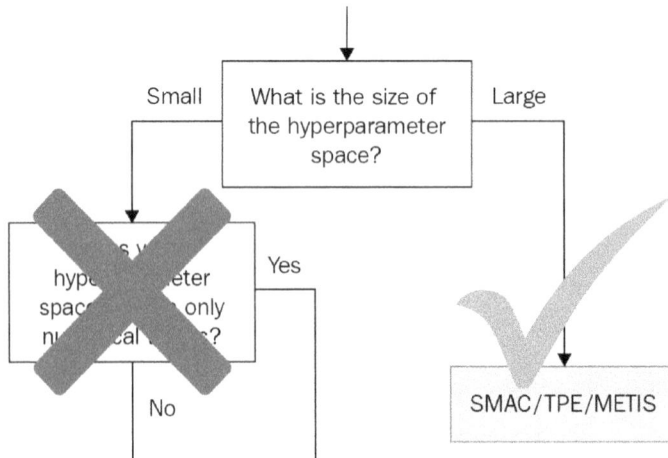

Figure 12.12 – Case study 3, medium-sized hyperparameter space with only numerical values

Based on the preceding diagram, we know that SMAC works best when the hyperparameter space is dominated by categorical hyperparameters, which is not the case here. Thus, we can remove SMAC from our options. If we try to decide based on the implementation popularity, then TPE is the one we should choose since it's implemented in Hyperopt, Optuna, and NNI, while Metis is only implemented in NNI. However, one of the main selling points of Metis is its ability to suggest the set of hyperparameters we should test in our next trial. So, which method should we pick? It's up to you.

In this section, we discussed the third case study on how to utilize HTDM in practice. Now, let's summarize this chapter.

Summary

In this chapter, we summarized all of the hyperparameter tuning methods we've discussed so far in a simple decision map called HTDM. This can help you to choose which method is the most suitable for your specific problem. We also discussed several important notes for each of the hyperparameter tuning methods and saw how to utilize the HTDM in practice. From now on, you'll be able to utilize HTDM in practice to help you decide which hyperparameter tuning method to adopt in your specific situation.

In the next chapter, we'll discuss the need to track hyperparameter tuning experiments and learn how to do so using several open source packages.

13

Tracking Hyperparameter Tuning Experiments

Working with a lot of experiments can sometimes be overwhelming. Many iterations of experiments will need to be done. It will become even more complicated when we are experimenting with many ML models.

In this chapter, you will be introduced to the importance of tracking hyperparameter tuning experiments, along with the usual practices. You will also be introduced to several open source packages that are available and learn how to utilize each of them in practice.

By the end of this chapter, you will be able to utilize your favorite package to track your hyperparameter tuning experiment. Being able to track your hyperparameter tuning experiment will boost the effectiveness of your workflow.

In this chapter, we will cover the following topics:

- Revisiting the usual practices
- Exploring Neptune
- Exploring Scikit-Optimize
- Exploring Optuna
- Exploring Microsoft NNI
- Exploring MLflow

Technical requirements

In this chapter, we will learn how to track hyperparameter tuning experiments with various packages. To ensure that you can reproduce the code examples in this chapter, you will require the following:

- The Python 3 (version 3.7 or above)
- The `pandas` package (version 1.3.4 or above)
- The `NumPy` package (version 1.21.2 or above)
- The `scikit-learn` package (version 1.0.1 or above)
- The `matplotlib` package (version 3.5.0 or above)
- The `Plotly` package (version 4.0.0 or above)
- The `Neptune-client` package (version 0.16.3 or above)
- The `Neptune-optuna` package (version 0.9.14 or above)
- The Scikit-Optimize package (version 0.9.0 or above)
- The `TensorFlow` package (version 2.4.1 or above)
- The `Optuna` package (version 2.10.0 or above)
- The `MLflow` package (version 1.27.0 or above)

All the code examples for this chapter can be found on GitHub at `https://github.com/PacktPublishing/Hyperparameter-Tuning-with-Python`.

Revisiting the usual practices

Conducting hyperparameter tuning experiments in a small-scale project may seem straightforward. We can easily do several iterations of experiments and write all the results in a separate document. We can log the details of the best set of hyperparameter values (or the tested set of hyperparameters if we perform a manual search method, as shown in *Chapter 3, Exhaustive Search*), along with the evaluation metric, in each experiment iteration. By having an experiment log, we can learn from the history and define a better hyperparameter space in the next iteration of the experiment.

When we adopt the automated hyperparameter tuning method (all the methods we've discussed so far besides the manual search method), we can get the final best set of hyperparameter values directly. However, this is not the case when we adopt the manual search method. We need to test numerous sets of hyperparameters manually. Several practices are adopted by the community when performing manual searches. Let's take a look.

Using a built-in Python dictionary

This is the most straightforward approach since we just need to create a Python dictionary that stores all the hyperparameter values that need to be tested. Although this practice is very simple, it has drawbacks. For example, we may not notice if we overwrite some of the hyperparameter values and forget to log the correct set of hyperparameter values. The following example of utilizing a built-in Python dictionary to store all of the hyperparameter values needs to be tested in a particular manual search iteration:

```
hyperparameters = {
 'n_estimators': 30,
 'max_features': 10,
 'criterion': 'gini',
 'max_depth': 5,
 'min_samples_split': 0.03,
 'min_samples_leaf': 1,
 }
```

Next, let's look at configuration files.

Using a configuration file

Whether it is a JSON, YAML, or CFG file, configuration files are another option. We can put all the hyperparameter details within this configuration file, along with other additional information, including (but not limited to) project name, author name, and data pre-processing pipeline methods. Once you have created the configuration file, you can load it into your Python script or Jupyter notebook, and treat it like a standard Python dictionary. The main advantage of using a configuration file is that all the important parameters are located within a single file, so it will be very easy to reuse the previously saved configuration files and increase the readability of your code. However, utilizing configuration files when working with a big project or huge code base can sometimes confuse us since we have to maintain several configuration files.

Using additional modules

The `argparse` and `Click` modules come in handy if you want to specify the hyperparameter values or any other training arguments via the **Command Line Interface** (**CLI**). These modules can be utilized when we write our code in a Python script, not in a Jupyter notebook.

Using argparse

The following code shows how to utilize `argparse` in a Python script:

```
import argparse
parser = argparse.ArgumentParser(description='Hyperparameter
Tuning')
parser.add_argument('--n_estimators, type=int, default=30,
help='number of estimators')
parser.add_argument('--max_features, type=int, default=20,
help='number of randomly sampled features for choosing the best
splitting point')
parser.add_argument('--criterion, type=str, default='gini',
help='homogeneity measurement method')
parser.add_argument('--max_depth, type=int, default=5,
help='maximum tree depth')
parser.add_argument('--min_samples_split, type=float,
default=0.03, help='minimum samples to split internal node')
parser.add_argument('--min_samples_leaf, type=int, default=1,
help='minimum number of samples in a leaf node')
parser.add_argument('--data_dir, type=str, required=True,
help='maximum tree depth')
```

The following code shows how to access the values from the CLI:

```
args = parser.parse_args()
print(args.n_estimators)
print(args.max_features)
print(args.criterion)
print(args.max_depth)
print(args.min_samples_split)
print(args.min_samples_leaf)
print(args.data_dir)
```

You can run the Python script with specified parameters, as follows:

```
python main.py --n_estimators 35 --criterion "entropy" --data_
dir "/path/to/my/data"
```

It is worth noting that the default values of the hyperparameters will be used if you don't specify them when calling the Python script.

Using click

The following code shows how to utilize `click` in a Python script. Note that `click` is very similar to `argparse` with a simpler implementation. We just need to add decorations on top of a particular function:

```python
import click
@click.command()
@click.option("--n_estimators, type=int, default=30,
help='number of estimators")
@click.option("--max_features, type=int, default=20,
help='number of randomly sampled features for choosing the best
splitting point")
@click.option("--criterion, type=str, default='gini',
help='homogeneity measurement method")
@click.option("--max_depth, type=int, default=5, help='maximum
tree depth")
@click.option("--min_samples_split, type=float, default=0.03,
help='minimum samples to split internal node")
@click.option("--data_dir, type=str, required=True,
help='maximum tree depth")
def hyperparameter_tuning(n_estimators, max_features,
criterion, max_depth, min_samples_split, data_dir):
#write your code here
```

Similar to `argparse`, you can run the Python script with specified parameters, as shown here. The default hyperparameter values will be used if you don't specify them when calling the Python script:

```
python main.py --n_estimators 35 --criterion "entropy" --data_
dir "/path/to/my/data"
```

While experimenting with either `argparse` or `click` is very easy to do, it is worth noting that neither saves values anywhere. Hence, it requires extra effort to log all of the experimented hyperparameter values in each trial.

Regardless of whether we are adopting manual search or other automated hyperparameter tuning methods, it will require a lot of effort if we have to log the resulting experiment's details manually. It can be overwhelming, especially when we are working with larger-scale experiments where we have to test several different ML models, data pre-processing pipelines, and other experiment setups. That's why, in the coming sections, you will be introduced to several packages that can help you track your hyperparameter tuning experiments so that you have a more effective workflow.

Exploring Neptune

Neptune is a Python (and R) package that acts as a metadata store for MLOps. This package supports a lot of features for working with the model-building metadata. We can utilize Neptune for tracking our experiments, not only hyperparameter tuning experiments but also other model-building-related experiments. We can log, visualize, organize, and manage our experiments just by using a single package. Furthermore, it also supports model registry and live monitors our ML jobs.

Installing Neptune is very easy – you can just use `pip install neptune-client` or `conda install -c conda-forge neptune-client`. Once it has been installed, you need to sign up for an account to get the API token. Neptune is free for an individual plan within the quota limit, but you need to pay if you want to utilize Neptune for commercial team usage. Further information about registering yourself for Neptune can be found on their official website: `https://neptune.ai/register`.

Using Neptune to help track your hyperparameter tuning experiments is straightforward, as shown in the following steps:

1. Create a new project from your Neptune account's home page:

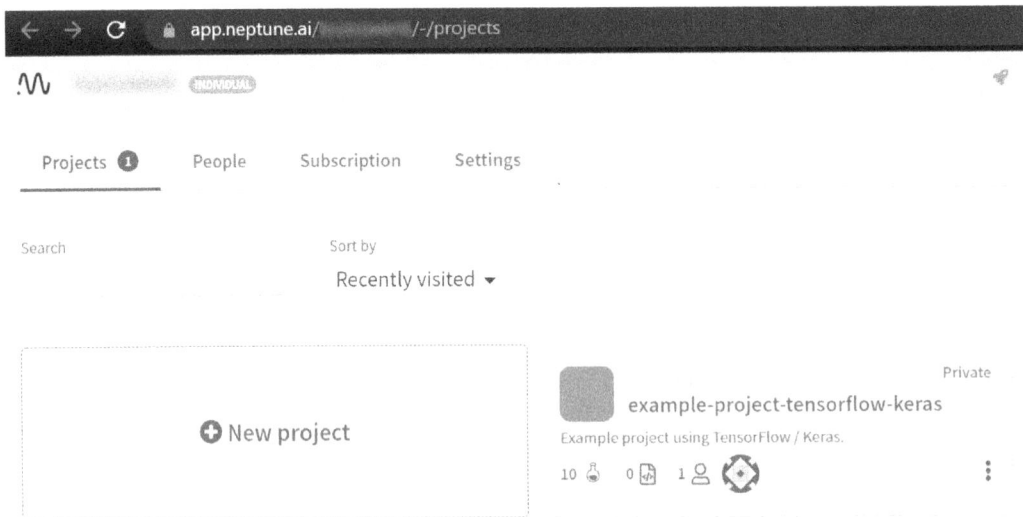

Figure 13.1 – Creating a new Neptune project

2. Enter a name and description for your project:

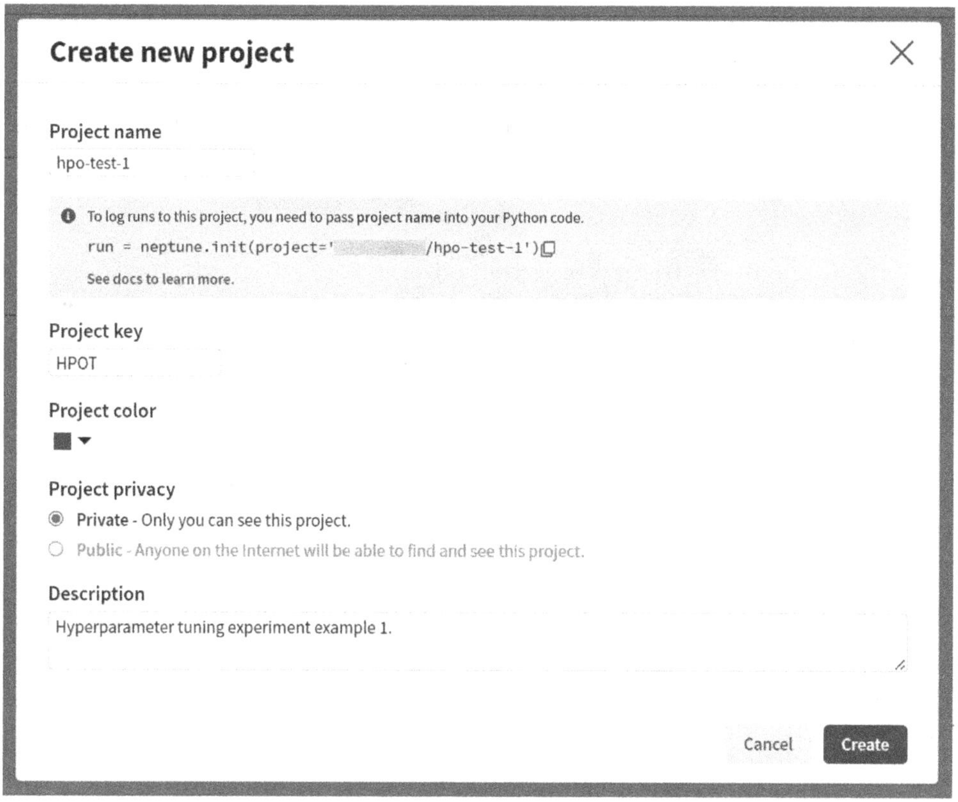

Figure 13.2 – Entering the project's details

3. Write the hyperparameter tuning experiment script. Neptune provides several boilerplate code options based on the framework you want to use, including (but not limited to) Optuna, PyTorch, Keras, TensorFlow, scikit-learn, and XGBoost. You can just copy the provided boilerplate code and customize it based on your needs. For example, let's use the provided boilerplate code for Optuna (see *Figure 13.3*) and save the training script as `train_optuna.py`. Please see the full code in this book's GitHub repository, which was provided in the *Technical requirements* section:

Figure 13.3 – Creating the hyperparameter tuning experiment script

4. Run the hyperparameter tuning script (`python train_optuna.py`) and look at the metadata of the experiments on your Neptune project page. Every run will be stored as a new experiment ID in Neptune, so you don't have to worry about the experiment versioning since Neptune will handle it automatically for you:

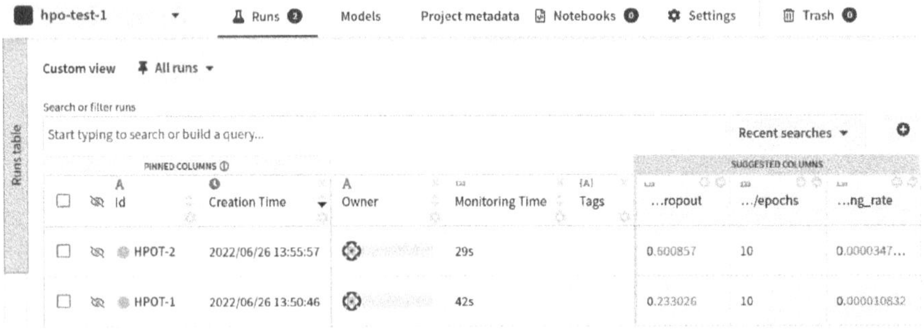

Figure 13.4 – Neptune's experiment runs table

You can also see all the metadata for each of the experiment runs, including (but not limited to) the tested hyperparameters, source code, CPU/GPU usage, metric charts, artifacts (data, model, or any other related files), and figures (for example, confusion matrices), as shown in the following screenshot:

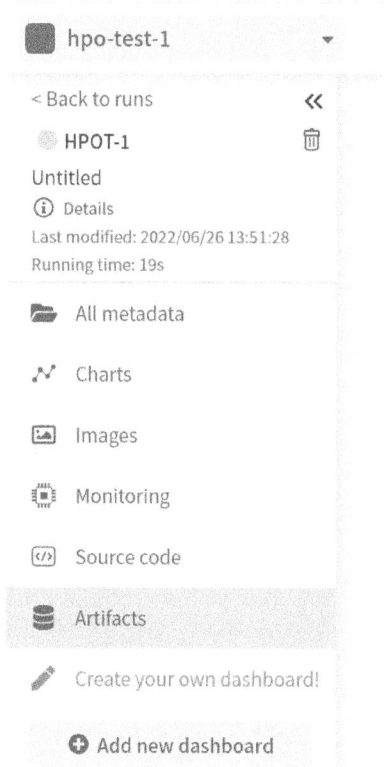

Figure 13.5 – Metadata stored in Neptune

5. Analyze the experiment results. Neptune can not only help you log all of the metadata for each experiment run, but it can also compare several different runs using several types of comparison strategies. You can see the hyperparameter values comparison via parallel plot or line charts. You can also compare all of the experiment details via a **Side-by-side** comparison strategy (see *Figure 13.6*). Furthermore, Neptune also enables us to compare the logged images or artifacts between each run:

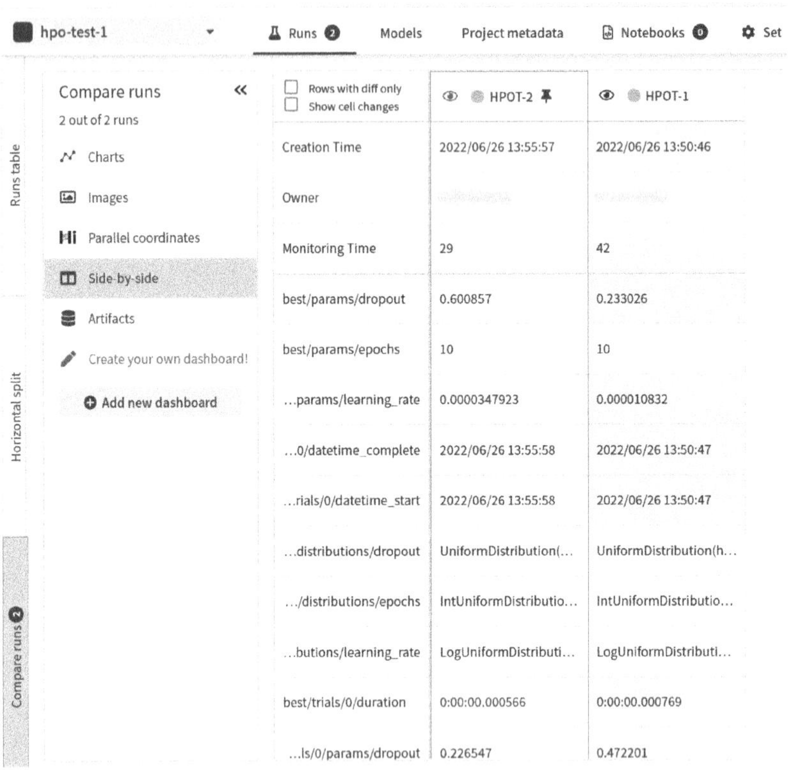

Figure 13.6 – Comparing the experiment runs and their results

For more information regarding what you can log and display in Neptune, please refer to the official documentation page: `https://docs.neptune.ai/you-should-know/what-can-you-log-and-display`.

Integrations in Neptune

Neptune provides numerous integrations for ML-related experiments in general and also for specific hyperparameter tuning-related tasks. Three integrations are supported by Neptune for hyperparameter tuning tasks: Optuna, Keras, and Scikit-Optimize. For more information, please refer to the official documentation page: `https://docs.neptune.ai/integrations-and-supported-tools/intro`.

> **More examples**
>
> Neptune is a very powerful package that can be utilized for other ML experiment-related tasks, too. For more examples of how to utilize Neptune in general, please refer to the official documentation page: `https://docs.neptune.ai/getting-started/examples`.

In this section, you were introduced to Neptune and how to utilize it to help you track your hyperparameter tuning experiments. In the next section, you will learn how to utilize the famous Scikit-Optimize package for hyperparameter tuning experiment tracking purposes.

Exploring scikit-optimize

You were introduced to the **Scikit-Optimize** package in *Chapter 7, Hyperparameter Tuning via Scikit*, to conduct a hyperparameter tuning experiment. In this section, we will learn how to utilize this package to track all hyperparameter tuning experiments conducted using this package.

Scikit-Optimize provides very nice visualization plots that summarize the tested hyperparameter values, the objective function scores, and the relationship between them. Three plots are available in this package, as shown here. Please see the full code in this book's GitHub repository for more details. The following plots were generated based on the same experimental setup that was provided in *Chapter 7, Hyperparameter Tuning via Scikit*, for the BOGP hyperparameter tuning method:

- `plot_convergence`: This is used to visualize the hyperparameter tuning optimization progress for each iteration:

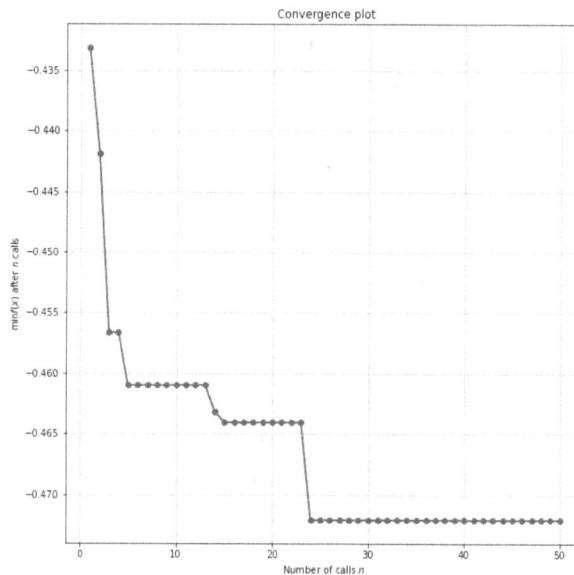

Figure 13.7 – Convergence plot

- `plot_evaluations`: This is used to visualize the optimization evolution process history. In other words, it shows the order in which hyperparameter values were sampled during the optimization process. For each hyperparameter, a histogram of explored hyperparameter values is generated. For each pair of hyperparameters, the scatter plot of tested hyperparameter values is visualized and equipped with colors to act as the legend of the evolution history (from blue to yellow):

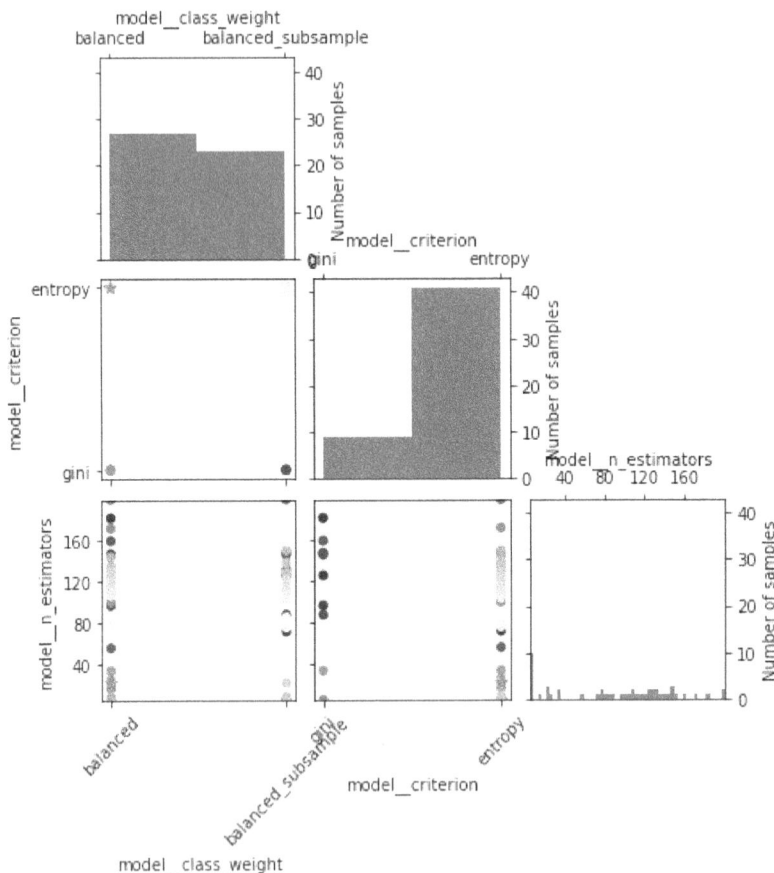

Figure 13.8 – Evaluation plot

- `plot_objective`: This is used to visualize the pairwise dependence plot of the objective function. This visualization helps us gain information regarding the relationship between the tested hyperparameter values and the objective function scores. From this plot, you can see which subspace needs more attention and which subspace, or even which hyperparameter, needs to be removed from the original space in the next trial:

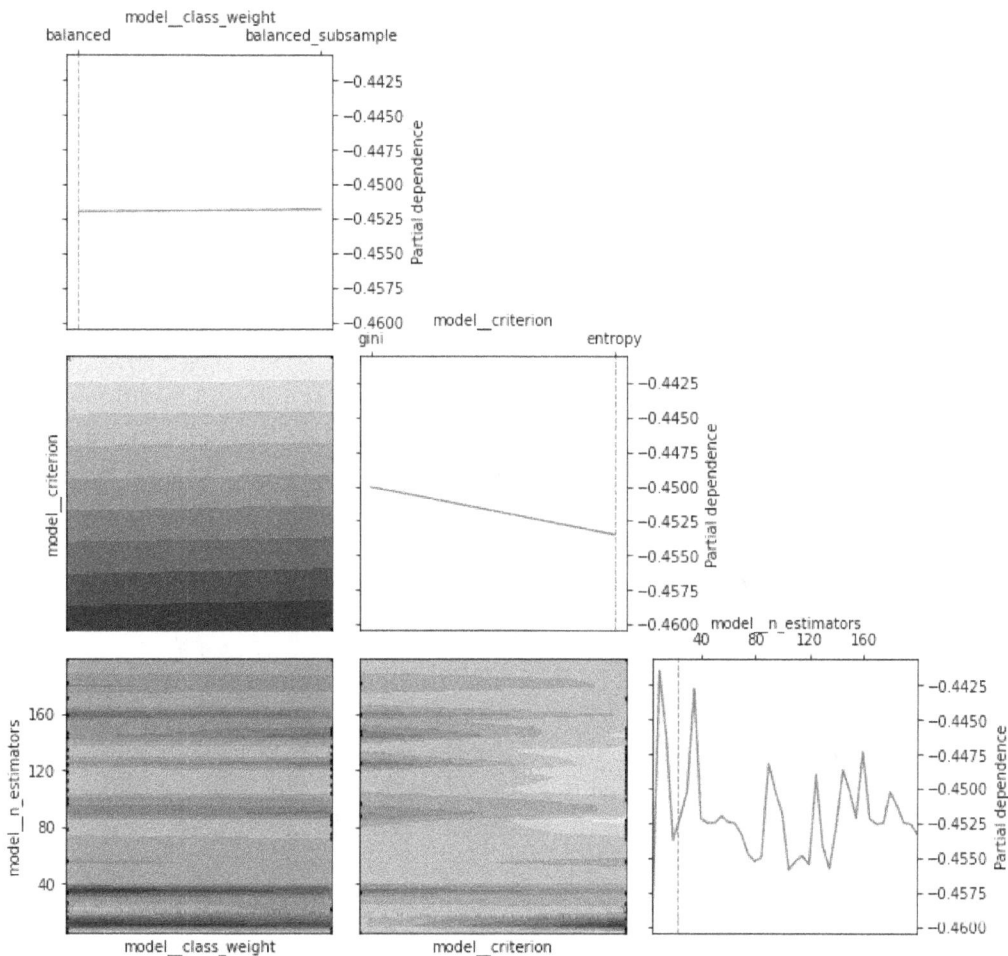

Figure 13.9 – Pairwise dependence plot

Integration with Neptune

Scikit-Optimize provides very informative visualization modules. However, it does not support any experiment versioning capabilities, unlike the Neptune package. To get the best of both worlds, we can integrate Scikit-Optimize with Neptune via its integration module. For more information about this, please refer to the official documentation page: `https://docs-legacy.neptune.ai/integrations/skopt.html`.

In this section, you learned how to utilize the Scikit-Optimize package to help you track your hyperparameter tuning experiments. In the next section, you will learn how to utilize the Optuna package for hyperparameter tuning experiment tracking purposes.

Exploring Optuna

Optuna is a hyperparameter tuning package in Python that provides several hyperparameter tuning methods. We discussed how to utilize Optuna to conduct a hyperparameter tuning experiment in *Chapter 9, Hyperparameter Tuning via Optuna*. Here, we will discuss how to utilize this package to track those experiments.

Similar to Scikit-Optimize, Optuna provides very nice visualization modules to help us track the hyperparameter tuning experiments and as a guide for us to decide which subspace to search in the next trial. Four visualization modules can be utilized, as shown here. All of them expect the `study` object (see *Chapter 9, Hyperparameter Tuning via Optuna*) as input. Please see the full code in this book's GitHub repository:

- `plot_contour`: This is used to visualize the relationship between hyperparameters (as well as the objective function scores) in the form of contour plots:

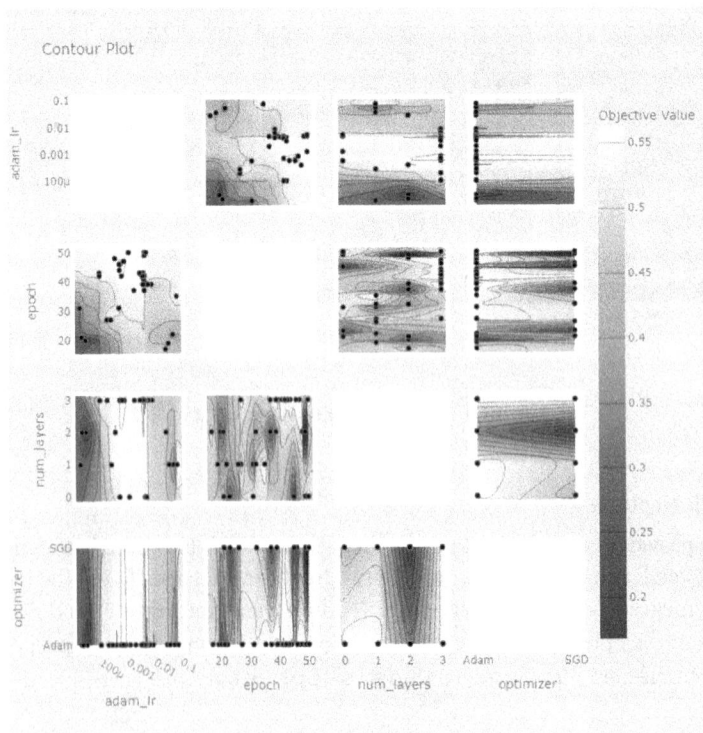

Figure 13.10 – Contour plot

- `plot_optimization_history`: This is used to visualize the hyperparameter tuning optimization progress for each iteration:

Optimization History Plot

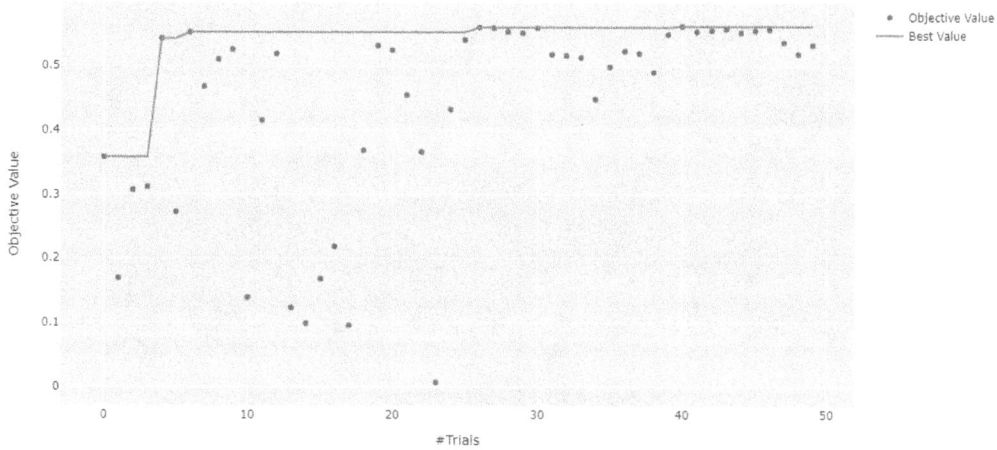

Figure 13.11 – Optimization history plot

- `plot_parallel_coordinate`: This is used to visualize the relationship between hyperparameters (as well as the objective function scores) in the form of a parallel coordinate plot:

Parallel Coordinate Plot

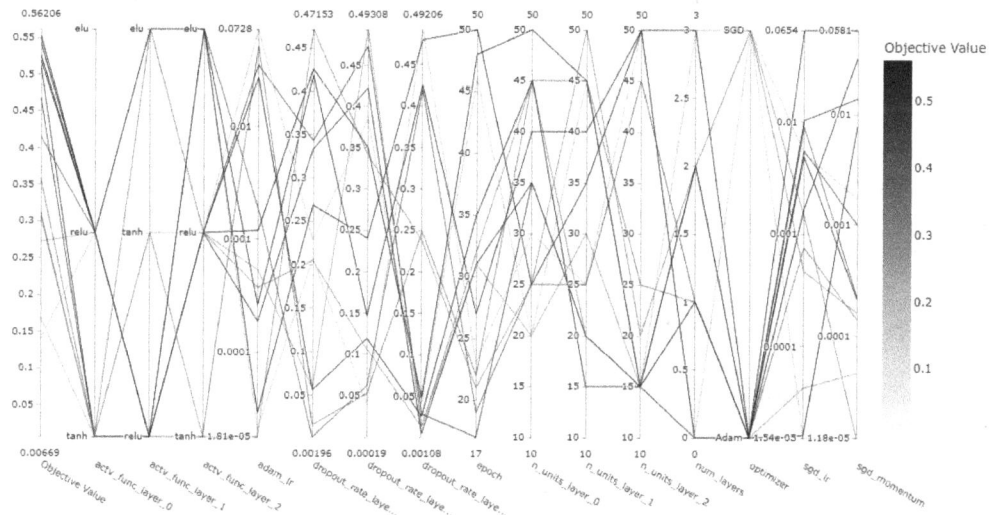

Figure 13.12 – Parallel coordinate plot

- `plot_slice`: This is used to visualize the hyperparameter tuning method's search evolution. You can see what hyperparameter values have been tested in the experiment and which subspace is getting more attention during the search process:

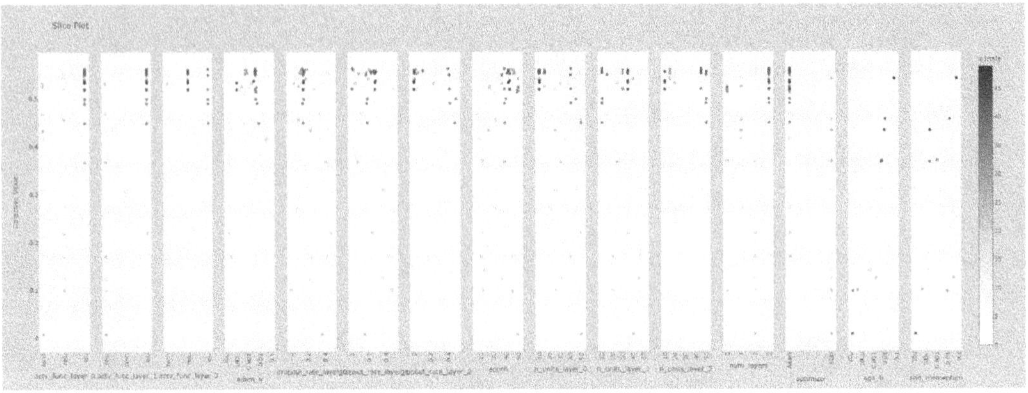

Figure 13.13 – Slice plot

The nice thing about all the visualization modules in Optuna is that they are all interactive charts since they are created using the `Plotly` visualization package. You can zoom in on a specific area in the charts and use other interactive features.

Integration with Neptune

Similar to Scikit-Optimize, Optuna provides very informative visualization modules. However, it does not support any experiment versioning capability, unlike the Neptune package. We can integrate Optuna with Neptune via its integration module. For more information about this, please refer to the official documentation page: `https://docs-legacy.neptune.ai/integrations/optuna.html`.

In this section, you learned how to utilize the Optuna package to track your hyperparameter tuning experiments. In the next section, you will learn how to utilize the Microsoft NNI package for hyperparameter tuning experiment tracking purposes.

Exploring Microsoft NNI

Neural Network Intelligence (**NNI**) is a package that is developed by Microsoft and can be utilized not only for hyperparameter tuning tasks but also for neural architecture search, model compression, and feature engineering. We discussed how to utilize NNI to conduct hyperparameter tuning experiments in *Chapter 10*, *Advanced Hyperparameter Tuning with DEAP and Microsoft NNI*.

In this section, we will discuss how to utilize this package to track those experiments. All of the experiment tracking modules provided by NNI are located in the *web portal*. You learned about the web portal in *Chapter 10, Advanced Hyperparameter Tuning with DEAP and Microsoft NNI*. However, we haven't discussed it in depth and there are many useful features you should know about.

The web portal can be utilized to visualize all of the hyperparameter tuning experiment's metadata, including (but not limited to) tuning and training progress, evaluation metrics, and error logs. It can also be utilized to update the experiment's concurrency and duration, and retry the failed trials. The following is a list of all the important modules in the NNI web portal that can be utilized to help us track our hyperparameter tuning experiments. The following plots have been generated based on the same experimental setup that was stated in *Chapter 10, Advanced Hyperparameter Tuning with DEAP and Microsoft NNI*, for the Random Search method. Please see the full code in this book's GitHub repository:

- The **Overview** page shows an overview of our hyperparameter tuning experiment, including its name and ID, status, start and end time, best metric, elapsed duration, number of trials faceted by the status, as well as the experiment path, training platform, and tuner details. Here, you can also change the maximum duration, the maximum number of trials, and the experiment's concurrency. There is also a dedicated module that shows the top-performing trials:

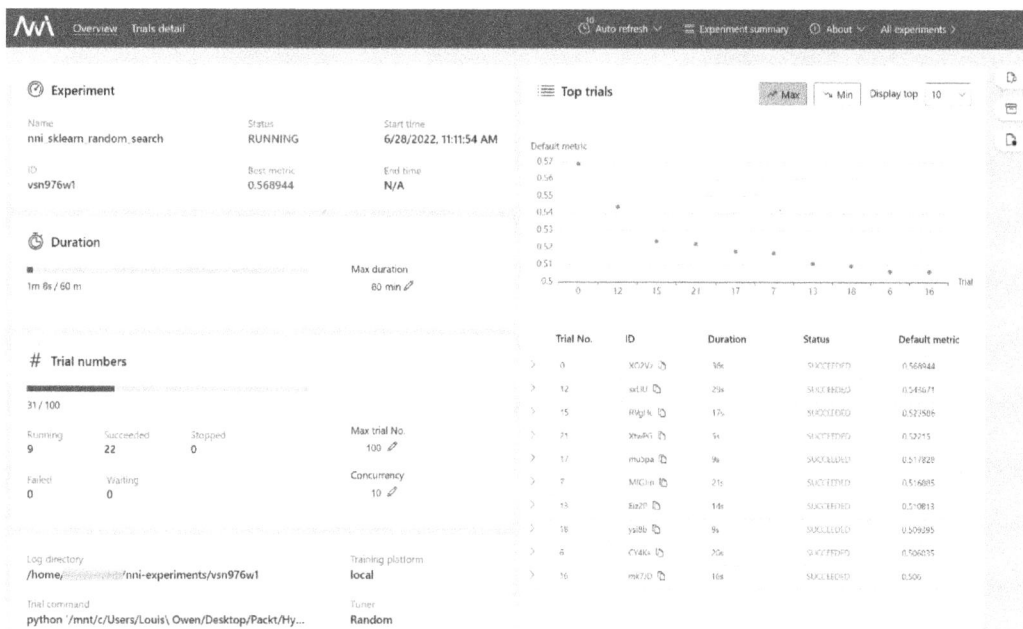

Figure 13.14 – The Overview page

- The **Trials detail** page shows every detail regarding the experiment's trials, including a visualization of all the metrics (see *Figure 13.15*), a hyperparameter values parallel plot (see *Figure 13.16*), a bar chart of the duration of all the trials (see *Figure 13.17*), and a line chart of all intermediate results that shows the trend of each trial during the intermediate steps. We can also see the details of each trial via the **Trial jobs** module, including (but not limited to) the trial's ID, duration, status, metric, hyperparameter value details, and log files (see *Figure 13.18*):

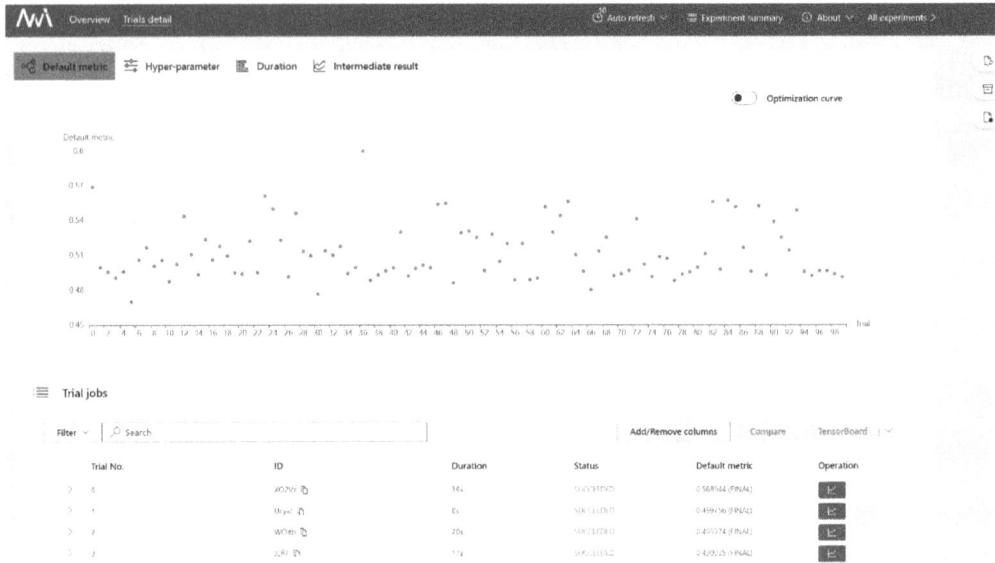

Figure 13.15 – The Trials detail page

The following is a parallel plot that shows different hyperparameter values that had been tested in the experiment:

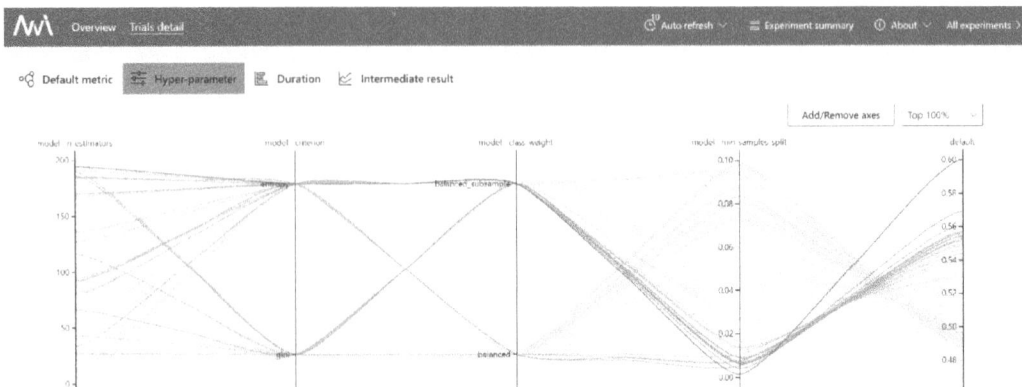

Figure 13.16 – Hyperparameter values parallel plot

The following is a bar chart containing information about the duration of all the trials in the experiment:

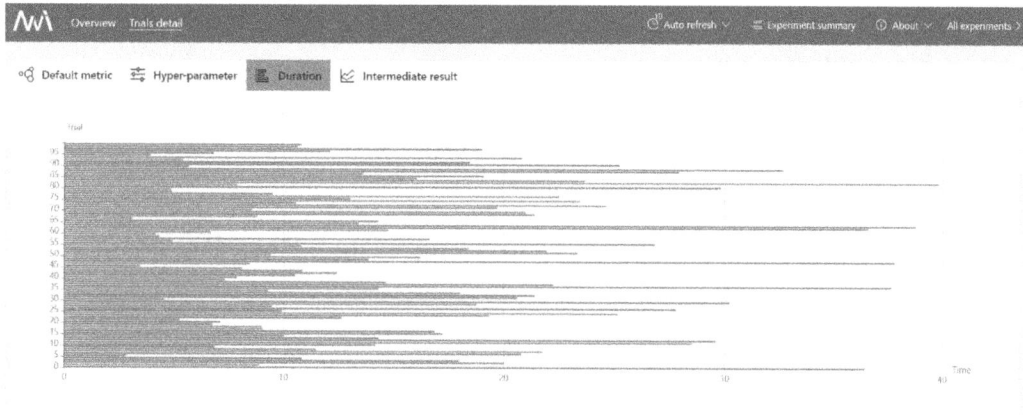

Figure 13.17 – Trials duration bar chart

Finally, there's the **Trial jobs** module:

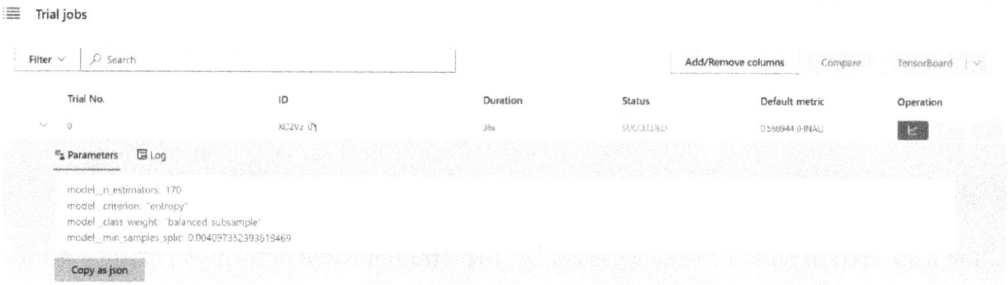

Figure 13.18 – The Trial jobs module

The Trial jobs module includes the following:

- **Sidebar**: We can access all the information related to the search space, config, and log files in the sidebar:

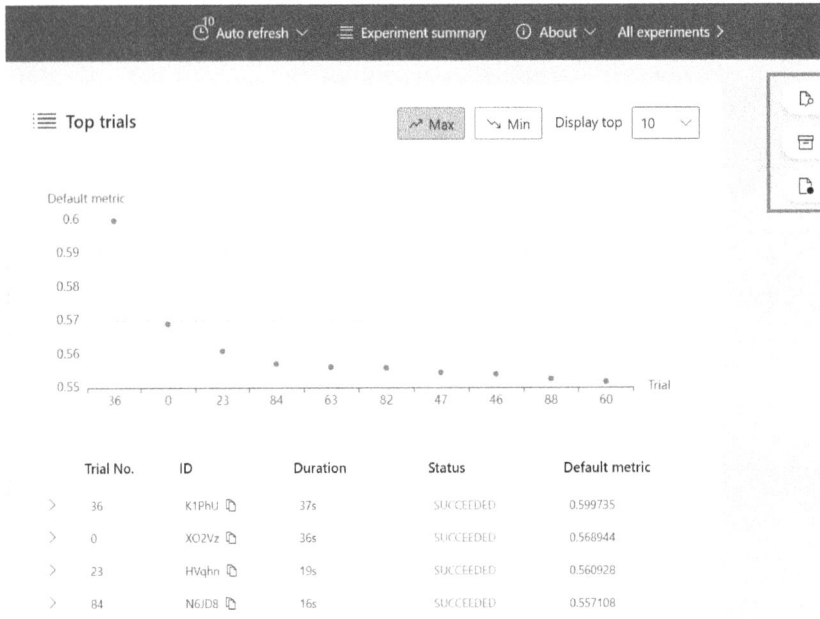

Figure 13.19 – Sidebar

- The **Auto refresh** button: We can also change the refresh interval of the web portal by using the **Auto refresh** button:

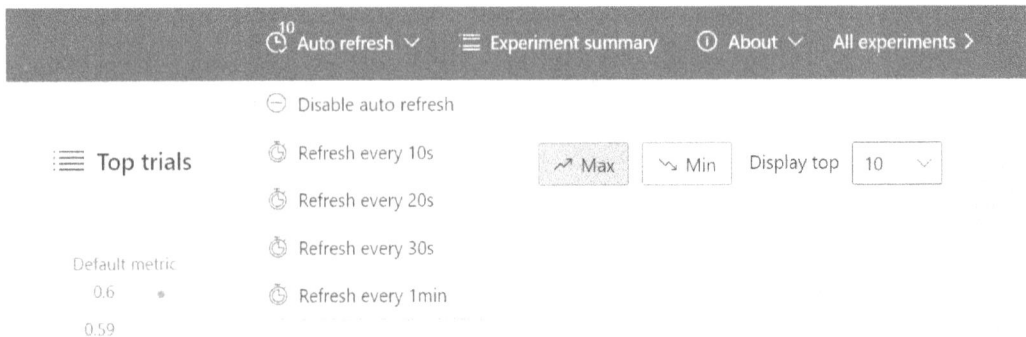

Figure 13.20 – The Auto refresh button

- The **Experiment summary** button: By clicking this button, you can view all the summaries for the current experiment:

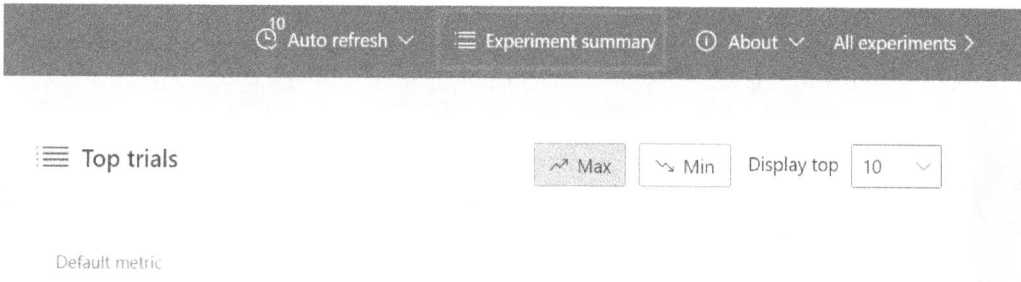

Figure 13.21 – The Experiment summary button

In this section, you learned how to utilize the Microsoft NNI package to track your hyperparameter tuning experiments. In the next section, you will learn how to utilize the MLflow package for hyperparameter-tuning experiment tracking purposes.

Exploring MLflow

MLflow can be utilized to manage the whole end-to-end ML pipeline. It is available in Python, R, Java, and via the REST API. The primary functions of MLflow include experiment tracking, ML code packaging, ML model deployment management, and centralized model storing and versioning. In this section, we will learn how to utilize this package to track our hyperparameter tuning experiments. Installing MLflow is very easy; you can just use the `pip install mlflow` command.

To track our hyperparameter tuning experiments with MLflow, we simply need to add several logging functions to our code base. Once we've added the required logging function, we can go to the provided UI by simply entering the `mlflow ui` command in the command line and opening it at `http://localhost:5000`. Many logging functions are provided by MLflow, and the following are some of the main important logging functions you need to be aware of. Please see the full example c

ode in this book's GitHub repository:

- `create_experiment()`: This function is used to create a new experiment. You can specify the name of the experiment, tags, and the path to store the experiment artifacts.

- `set_experiment()`: This function is used to set the given experiment name or ID as the current active experiment.

- `start_run()`: This function is used to start a new MLflow run under the current active experiment. It is suggested to use this function as a context manager within a `with` block.

- `log_metric()`: This function is used to log a single metric within the currently active run. If you want to do bulk logging, you can also use the `log_metrics()` function by passing a dictionary of metrics.

- `log_param()`: This function is used to log a parameter or hyperparameter within the currently active run. If you want to do bulk logging, you can also use the `log_params()` function by passing a dictionary of metrics.

- `log_artifact()`: This function is used to log a file or directory as an artifact of the currently active run. If you want to log all the contents of a local directory, you can also use the `log_artifacts()` function.

- `set_tag()`: This function is used to set a tag for the currently active run. You must provide the key and value of the tag. For example, you can set the key as `"release_version"` and the value as `"1.0.0"`.

- `log_figure()`: This function is used to log a figure as an artifact of the currently active run. This function supports the `matplotlib` and `pyplot` figure object types.

- `log_image()`: This function is used to log an image as an artifact of the currently active run. This function supports the `numpy.ndarray` and `PIL.image.image` object types.

MLflow Logging Functions

For more information regarding all the available logging functions in MLfLow, please refer to the official documentation page: `https://www.mlflow.org/docs/latest/tracking.html#logging-functions`.

MLflow Integrations

MLflow also supports integrations with many well-known open source packages, including (but not limited to) scikit-learn, TensorFlow, XGBoost, PyTorch, and Spark. You can do automatic logging by utilizing the provided integrations. For more information, please refer to the official documentation page: `https://www.mlflow.org/docs/latest/tracking.html#automatic-logging`.

Examples of Hyperparameter Tuning Use Cases

The author of MLflow has provided example code for hyperparameter tuning use cases. For more information, please refer to the official GitHub repository: `https://github.com/mlflow/mlflow/tree/master/examples/hyperparam`.

In this section, you learned how to utilize the MLflow package to track your hyperparameter tuning experiments. You can start exploring this package by yourself to get a better understanding of how this package works and how powerful it is.

Summary

In this chapter, we discussed the importance of tracking hyperparameter tuning experiments, along with the usual practices. You were also introduced to several open source packages that are available and learned how to utilize each of them in practice, including Neptune, Scikit-Optimize, Optuna, Microsoft NNI, and MLflow. At this point, you should be able to utilize your favorite package to track your hyperparameter tuning experiment, which will boost the effectiveness of your workflow.

In the next chapter, we'll conclude all the topics we have discussed throughout this book. We'll also discuss the next steps you can take to expand your hyperparameter tuning knowledge.

Conclusions and Next Steps

Congratulations on finishing this book! You have been introduced to a lot of interesting concepts, methods, and implementations related to hyperparameter tuning throughout the previous chapters. This chapter summarizes all the important lessons learned in the previous chapters, and will introduce you to several topics or implementations that you may benefit from that we have not covered yet in this book.

The following are the main topics that will be discussed in this chapter:

- Revisiting hyperparameter tuning methods and packages
- Revisiting HTDM
- What's next?

Revisiting hyperparameter tuning methods and packages

Throughout this book, we have discussed four groups of hyperparameter tuning methods, including exhaustive search, Bayesian optimization, heuristic search, and multi-fidelity optimization. All the methods within each group have similar characteristics to each other. For example, manual search, grid search, and random search, which are part of the exhaustive search group, all work by exhaustively searching through the hyperparameter space, and can be categorized as uninformed search methods.

Bayesian optimization hyperparameter tuning methods are categorized as informed search methods, where all of them work by utilizing both surrogate model and acquisition function. Hyperparameter tuning methods, which are part of the heuristic search group, work by performing trial and error. As for hyperparameter tuning methods from the multi-fidelity optimization group, they all utilize the cheap approximation of the whole hyperparameter tuning pipeline, so that we can have similar performance results with much lesser computational cost and faster experiment time.

The following table summarizes all of the hyperparameter tuning methods discussed in this book, along with the supported packages:

Group	Method	Notes	Supported Packages
Exhaustive Search	Manual Search	Best combined with other methods.	
	Grid Search	- Know exactly what the important hyperparameters to be tuned are. - Want to get an exact optimal solution.	scikit-learn, Optuna, and NNI
	Random Search	Works better than grid search when there's an unimportant hyperparameter in the space.	scikit-learn, Hyperopt, Optuna, and NNI
Bayesian Optimization	BOGP	There are variants of BOGP that also support non-numerical types of hyperparameters.	scikit-learn and NNI
	SMAC	Best when the hyperparameter space is dominated by categorical hyperparameters.	NNI
	TPE	Unlike SMAC, TPE is not focusing only on the best-observed points during the trials, but it focuses on the distribution of the best-observed points instead.	Hyperopt, Optuna, and NNI
	Metis	Like SMAC and TPE, but also want to know which set of hyperparameters should be tested in the next trial.	NNI
Heuristic Search	Simulated Annealing	May skip parts of the search space that contain optimal hyperparameters.	Hyperopt, Optuna, and NNI
	Genetic Algorithm	High computation cost due to the need to evaluate all individuals in each generation.	DEAP
	PSO	Works well only with a continuous type of hyperparameter but can be modified for discrete hyperparameters as well.	DEAP
	PBT	Just want the final trained model without knowing the hyperparameter configuration.	NNI

Group	Method	Notes	Supported Packages
Multi-Fidelity Optimization	Coarse-to-Fine	A very simple method with a customizable module based on your own preference.	scikit-learn
	Successive Halving	Has metadata or previous experience setting up the trade-off between the number of resources and candidates to be sampled.	scikit-learn and Optuna
	Hyper Band	Like Successive Halving but does not have time or metadata to help you configure the trade-off between the number of resources and candidates.	scikit-learn, Optuna, and NNI
	BOHB	- Able to decide which subspace needs to be searched based on previous experiences, not based on luck. - Not only get a strong initial performance (inherited from HB) but also a strong final performance (inherited from BO).	NNI

Figure 14.1 – Hyperparameter tuning methods and packages summary

In this section, we have revisited all of the hyperparameter tuning methods and packages discussed throughout the book. In the next section, we will revisit the HTDM.

Revisiting HTDM

The **Hyperparameter Tuning Decision Map (HTDM)** is a map that you can use to help you decide which hyperparameter tuning method should be adopted in a particular situation. We discussed in detail how you can utilize HTDM, along with several use cases, in *Chapter 12, Introducing the Hyperparameter Tuning Decision Map*. Here, we will only revisit the map, as shown in the following figure:

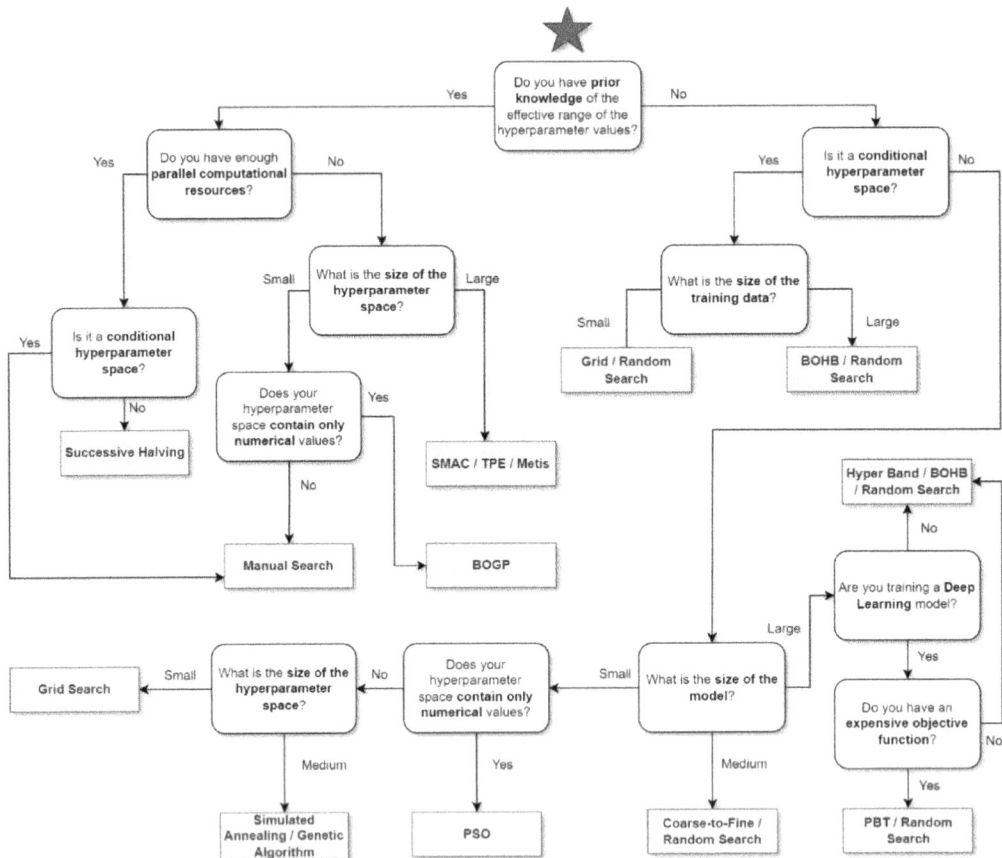

Figure 14.2 – HTDM

In this section, we have revisited the HTDM. In the next section, we'll discuss other topics you may find interesting to further boost your hyperparameter tuning knowledge.

What's next?

Even though we have discussed a lot of hyperparameter tuning methods and their implementations in various packages, there are several important concepts you may need to know about that have not been discussed in this book. As for the hyperparameter tuning method, you can also read more about the **CMA-ES** method, which is part of the heuristic search group (`https://cma-es.github.io/`). You can also read more about the **meta-learning** concept to further boost the performance of your Bayesian optimization tuning results (`https://lilianweng.github.io/posts/2018-11-30-meta-learning/`). It is also worth noting that we can combine the manual search method with other hyperparameter tuning methods to boost the efficiency of our experiments, especially when we already have prior knowledge about the good range of the hyperparameter values.

As for the packages, you can also learn more about the **HpBandSter** package, which implements the Hyper Band, BOHB, and random search methods (`https://github.com/automl/HpBandSter`). Finally, there are also several packages that automatically create a scikit-learn wrapper from the non-scikit-learn model. For example, you can utilize the **Skorch** package to create scikit-learn wrappers from PyTorch models (`https://skorch.readthedocs.io/en/stable/`).

Summary

In this chapter, we have summarized all the important concepts discussed throughout all chapters in this book. You have also been introduced to several new concepts that you may want to learn to further boost your hyperparameter tuning knowledge. From now on, you will have the skills you need to take full control over your machine learning models and get the best models for the best results via hyperparameter tuning experiments.

Thanks for investing your interest and time in reading this book. Best of luck on your hyperparameter tuning learning journey!

Index

Packt>

Packt.com

Subscribe to our online digital library for full access to over 7,000 books and videos, as well as industry leading tools to help you plan your personal development and advance your career. For more information, please visit our website.

Why subscribe?

- Spend less time learning and more time coding with practical eBooks and Videos from over 4,000 industry professionals

- Improve your learning with Skill Plans built especially for you

- Get a free eBook or video every month

- Fully searchable for easy access to vital information

- Copy and paste, print, and bookmark content

Did you know that Packt offers eBook versions of every book published, with PDF and ePub files available? You can upgrade to the eBook version at packt.com and as a print book customer, you are entitled to a discount on the eBook copy. Get in touch with us at customercare@packtpub.com for more details.

At www.packt.com, you can also read a collection of free technical articles, sign up for a range of free newsletters, and receive exclusive discounts and offers on Packt books and eBooks.

Other Books You May Enjoy

If you enjoyed this book, you may be interested in these other books by Packt:

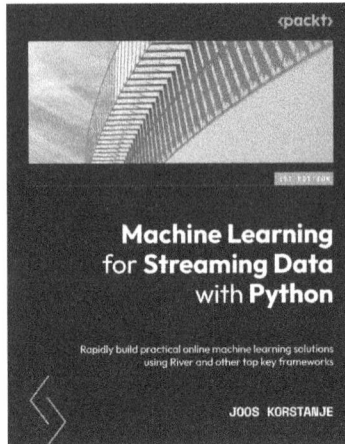

Machine Learning for Streaming Data with Python

Joos Korstanje

ISBN: 9781803248363

- Understand the challenges and advantages of working with streaming data Develop real-time insights from streaming data Understand the implementation of streaming data with various use cases to boost your knowledge Develop a PCA alternative that can work on real-time data Explore best practices for handling streaming data that you absolutely need to remember Develop an API for real-time machine learning inference

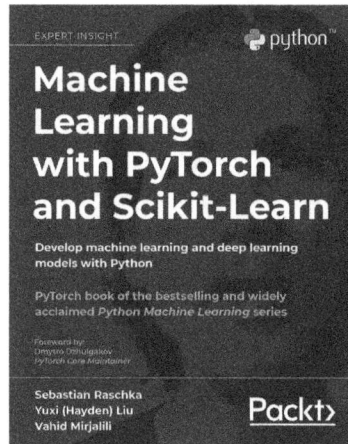

Machine Learning with PyTorch and Scikit-Learn

Sebastian Raschka, Yuxi (Hayden) Liu, Vahid Mirjalili

ISBN: 9781801819312

- Explore frameworks, models, and techniques for machines to 'learn' from data Use scikit-learn for machine learning and PyTorch for deep learning Train machine learning classifiers on images, text, and more Build and train neural networks, transformers, and boosting algorithms Discover best practices for evaluating and tuning models Predict continuous target outcomes using regression analysis Dig deeper into textual and social media data using sentiment analysis

Packt is searching for authors like you

If you're interested in becoming an author for Packt, please visit `authors.packtpub.com` and apply today. We have worked with thousands of developers and tech professionals, just like you, to help them share their insight with the global tech community. You can make a general application, apply for a specific hot topic that we are recruiting an author for, or submit your own idea.

Share Your Thoughts

Now you've finished *Hyperparameter Tuning with Python*, we'd love to hear your thoughts! Scan the QR code below to go straight to the Amazon review page for this book and share your feedback or leave a review on the site that you purchased it from.

`https://packt.link/r/1-803-23587-X`

Your review is important to us and the tech community and will help us make sure we're delivering excellent quality content.